AF361912

Physical Activity and Sports Practice in Improving Body Composition and Sustainable Health

Physical Activity and Sports Practice in Improving Body Composition and Sustainable Health

Editors

Badicu Georgian
Francesco Campa

MDPI • Basel • Beijing • Wuhan • Barcelona • Belgrade • Manchester • Tokyo • Cluj • Tianjin

Editors
Badicu Georgian
Department of Physical
Education and Special Motricity,
Faculty of Physical Education
and Mountain Sports,
Transilvania University of Brasov
Romania

Francesco Campa
Department for Life Quality
Studies, University of Bologna
Italy

Editorial Office
MDPI
St. Alban-Anlage 66
4052 Basel, Switzerland

This is a reprint of articles from the Special Issue published online in the open access journal *Sustainability* (ISSN 2071-1050) (available at: https://www.mdpi.com/journal/sustainability/special_issues/phy_sport_health_sus).

For citation purposes, cite each article independently as indicated on the article page online and as indicated below:

LastName, A.A.; LastName, B.B.; LastName, C.C. Article Title. *Journal Name* **Year**, *Volume Number*, Page Range.

ISBN 978-3-0365-0598-5 (Hbk)
ISBN 978-3-0365-0599-2 (PDF)

Contents

About the Editors

Badicu Georgian is an associate professor at Transilvania University of Brasov, Faculty of Physical Education and Mountain Sports, Department of Physical Education and Special Motricity. He has published about 12 papers in JCR journals and is the author or co-author of 11 papers published by MDPI. His main interests are physical activity, fitness, education, obesity, well-being, recreation, football, public health, quality of life, sports activities, physical education, didactics of physical education, and sports.

Francesco Campa Adjunct professor and research fellow at the Department for Life Quality Studies at the University of Bologna (Italy) received his PhD in pharmacology and toxicology, human development and movement sciences from the University of Bologna. He has published one book on sports anthropometry, several book chapters, and over 40 peer-reviewed journal and conference papers (https://orcid.org/0000-0002-3028-7802), H-index 10. His main interests are sports anthropometry, bioimpedance vector analysis, and body composition optimization applied to sports performance.

Preface to "Physical Activity and Sports Practice in Improving Body Composition and Sustainable Health"

Analyzing and monitoring body composition and physical activity is an important topic when discussing the benefits of leading a healthy lifestyle, due to its influence on health status. Sports practice and a healthy lifestyle result in optimal body composition, therefore positively contributing to improvements in sports performance. In the last few years, the scope of research in sports has become increasingly wide and detailed, laying the foundations for the development of innovative evaluation approaches aimed at improving body composition, health, and physical function. The articles published in this research collection highlight the relationship among these specific parameters through longitudinal and transversal experimental designs as well as systematic literature reviews.

Badicu Georgian, Francesco Campa
Editors

 *sustainability*

Article

Functional Autonomy Evaluation Levels in Middle-Aged and Older Spanish Women: On Behalf of the Healthy-Age Network

Pablo Jorge Marcos-Pardo [1,2,*], Noelia González-Gálvez [1,2,*], Raquel Vaquero-Cristóbal [1,2], Gemma María Gea-García [1,2], Abraham López-Vivancos [1,2], Alejandro Espeso-García [1,2], Daniel Velázquez-Díaz [2,3,4], Ana Carbonell-Baeza [2,3,4], David Jiménez-Pavón [2,3,4], Juliana Brandão Pinto de Castro [5] and Rodrigo Gomes de Souza Vale [2,5,6]

[1] Research Group on Health, Physical Activity, Fitness and Motor Behaviour (GISAFFCOM) and Physical Activity and Sport Sciences Department, Faculty of Sport, Catholic University San Antonio of Murcia, 30107 Murcia, Spain; rvaquero@ucam.edu (R.V.-C.); gmgea@ucam.edu (G.M.G.-G.); alvivancos@ucam.edu (A.L.-V.); aespeso@ucam.edu (A.E.-G.)

[2] Active Aging, Exercise and Health/HEALTHY-AGE Network, Consejo Superior de Deportes (CSD), Ministry of Culture and Sport of Spain, 28040 Madrid, Spain; daniel.velazquez@uca.es (D.V.-D.); ana.carbonell@uca.es (A.C.-B.); david.jimenez@uca.es (D.J.-P.); rodrigovale@globo.com (R.G.d.S.V.)

[3] MOVE-IT Research Group, Department of Physical Education, Faculty of Education Sciences, University of Cádiz, 11519 Cádiz, Spain

[4] Biomedical Research and Innovation Institute of Cádiz (INiBICA) Research Unit, Puerta del Mar University Hospital, University of Cádiz, 11519 Cádiz, Spain

[5] Laboratory of Exercise and Sport (LABEES), Rio de Janeiro State University (UERJ), 20550 Rio de Janeiro, Brazil; julianabrandaoflp@hotmail.com

[6] Exercise Physiology Laboratory, Estacio de Sá University, 22640-102 Cabo Frio, Brazil

[*] Correspondence: pmarcos@ucam.edu (P.J.M.-P.); ngonzalez@ucam.edu (N.G.-G.)

Received: 30 September 2020; Accepted: 2 November 2020; Published: 5 November 2020

Abstract: Aging is associated with a progressive loss of functional capacity that affects the health and quality of life of middle-aged and older people. The purpose of this study was to report functional autonomy evaluation levels in middle-aged and older women in the Spanish context. A total of 709 middle-aged and older women, between 50 and 90 years old, were selected to participate in the study. The sample was divided by age category every five years. The functional autonomy levels were determined by the Latin American Group for Maturity (GDLAM) protocol and we developed a classification pattern for middle-aged and older women living in Spain. The GDLAM Index (GI) was then calculated to assess functional autonomy. The classification of the tests and the GI followed the percentile rank (P) Very Good ($p < 0.15$), Good (p 0.16–p 0.50), Regular (p 0.51–p 0.85), and Poor ($p > 0.85$). It was considered that the lower the value found for the percentile, the better the result. The GDLAM protocol showed strong reliability with intraclass correlation coefficient (ICC) values greater than 0.92 in all tests. It is observed that all variables of the GDLAM protocol presented a positive and significant correlation with age ($p < 0.001$). The Roc Curve showed that GI values higher than 26 (CI95% = 0.97–1.00; $p < 0.001$) and 32 (CI95% = 0.98–1.00; $p < 0.001$) for middle-aged and elderly women, respectively, can predict and indicate low functional autonomy. The normative values hereby provided will enable evaluation and adequate interpretation of Spanish middle-aged and older women's functional autonomy.

Keywords: functional capacity; older people; functional autonomy assessment; aging; health

---

1. Introduction

Aging is associated with a decrease in the efficiency of several physiological processes, including a progressive loss of functional capacity. The functional decrease tends to occur earlier in women, mainly around menopause [1], involving postural control instabilities, which lead to changes in walking and balance [2]. Besides balance, other components of physical fitness that influences health are muscle strength, aerobic capacity, and flexibility in adults as well in older people [3,4]. These physical capabilities are directly related to the performance of activities of daily living (ADL) and prevention and control of non-communicable diseases [3,5]. Thus, functional autonomy is one of the most relevant markers related to health, quality of life, and performance of ADL of adults and older people [6].

The Latin American Group for Maturity (GDLAM) defines autonomy from three perspectives: autonomy of action, which refers to physical independence; autonomy of will, which means the possibility of self-determination; autonomy of thoughts, which enables the individual to judge any situation. According to GDLAM, independence is the ability to perform tasks without help of people, devices, or systems [7]. The decline in functional autonomy is one of the main consequences of aging and can lead to frailty of older individuals [8,9]. Frailty occurs in a multifactorial manner as a result of different physiological regressive processes associated with aging [10]. This progressive deterioration, together with the decrease in strength and endurance, places the individual in a vulnerable condition of functional capacity and quality of life [11]. Quality of life is one of the factors responsible for the population's longevity. Maintaining lifestyle habits that promote functional autonomy in older individuals requires an awareness effort. Hence, physical activities are important to achieve the desired standard in certain features of quality of life and functional autonomy in these individuals [12]. Specifically in Spain, population projections estimate that, in the coming decades, the population aged 65 and over will continue to increase. The projection of older people for 2060 is one-third of the total Spanish population, which corresponds to over 16 million older people, being women are in the majority, outnumbering men by 32% [13]. Furthermore, in recent years, the physical inactivity levels of Spanish young populations have increased [14], which brings a red alert as age-related declines in functional capacity are clearly accentuated by physical inactivity [15], thus these inactivity levels in the young increase the risk of dependence later in life [16].

Therefore, more research on ADL performance of older people, especially in women as an understudied population, can give health professionals precise standards to classify functional autonomy and verify exercise training program efficiency [17]. One instrument used in many studies [18–25] to evaluate the functional autonomy levels in older people is the GDLAM protocol. This protocol measures the time it takes, in seconds, to complete five typical ADL [17]. Then, the current study aims to determine the functional autonomy levels using the GDLAM protocol and to develop a classification pattern for middle-aged and older women living in Spain.

2. Materials and Methods

2.1. Participants

The study initially counted 1054 older adults from Social Care Programs for older people in Murcia and Cádiz, Spain, from April to November 2019. These individuals passed through the following inclusion criteria: being female; being 50 years of age or older; being independent in their ADL; could be doing physical exercise or not. Any type of acute or chronic condition that could compromise or become an impediment factor for the performance of functional autonomy tests was considered an exclusion criterion, such as: cardiopathies, diabetes, hypertension, and uncontrolled bronchitis as well as any musculoskeletal conditions that could be an interaction factor for the tests (osteoarthritis, recent fracture, tendinitis, and prosthesis use); neurological problems; morbid obesity; chronic renal individuals and those who used medication that could cause attention disorders.

After the end of the sampling process, 709 older women were selected to participate in the study. The participants, who ranged between 50 and 90 years old, were divided by age category every five years.

The survey participants signed the informed consent term according to the Helsinki Declaration [26]. The study was approved by the Ethics and Research Committee involving Human Beings of the Catholic University San Antonio of Murcia under protocol n. CE031907.

2.2. Data Collection Procedures

2.2.1. Anthropometric Evaluation

A mechanical scale with a capacity of 150 kg and a precision of 100 g, with a stadiometer, from Filizola (Brazil), was used for the evaluation of body mass and height, following the protocol of the International Society for the Advancement of Kinanthropometry [27] for body mass index (BMI) calculation.

2.2.2. Functional Autonomy Evaluation

Functional autonomy was assessed through the Latin American Group for Maturity (GDLAM) protocol of autonomy [7,17,25,28] composed of the following tests:

(1) Walk 10 m (W10 m): the purpose of this test is to evaluate the gait speed of the individual for a distance of 10 m [29] (Figure 1a).
(2) To sit and get up from the chair and move around the house (SCMA): the objective is to assess the ability of the middle-aged and older people in their agility and balance in life situations. With a chair fixed on the ground, two cones should be demarked diagonally to the chair, at a distance of four meters behind and three meters to the right and left sides of the same. The individual begins the test seated in the chair, with her feet on the floor, and with the sign "already," she gets up, goes to the right, moves around the cone, returns to the chair, sits down, and takes both feet off the ground. Without hesitating, she does the same move to the left. Immediately, the same course is again completed, to the right and left, thus making the entire journey and circulating each cone twice, in the shortest time possible [30] (Figure 1b).
(3) Stand up from sitting position (SSP): the test aims to assess the functional capacity of the lower limbs. The test starts with the individual in the sitting position in a chair without arm support, and the seat at a distance from the ground of 50 cm, then, the individual stands up and takes a seat five times consecutively [31] (Figure 1c).
(4) Standing up from the prone position (SPP): the purpose of this test is to assess the overall ability of the individual to get up the floor. The test starts with the individual in a ventral decubitus position, with arms along the body; the command of "now", indicates that the individual must get up, leaving the position as soon as possible [32] (Figure 1d).
(5) To put on and take off a T-shirt (PTS): the individual should be standing with arms along the body and a T-shirt in one of the hands. At the voice signal of "Go," the individual should put on the shirt and immediately take it off, returning to the starting position. This test is intended to measure the agility and coordination of the upper limb [33] (Figure 1e).

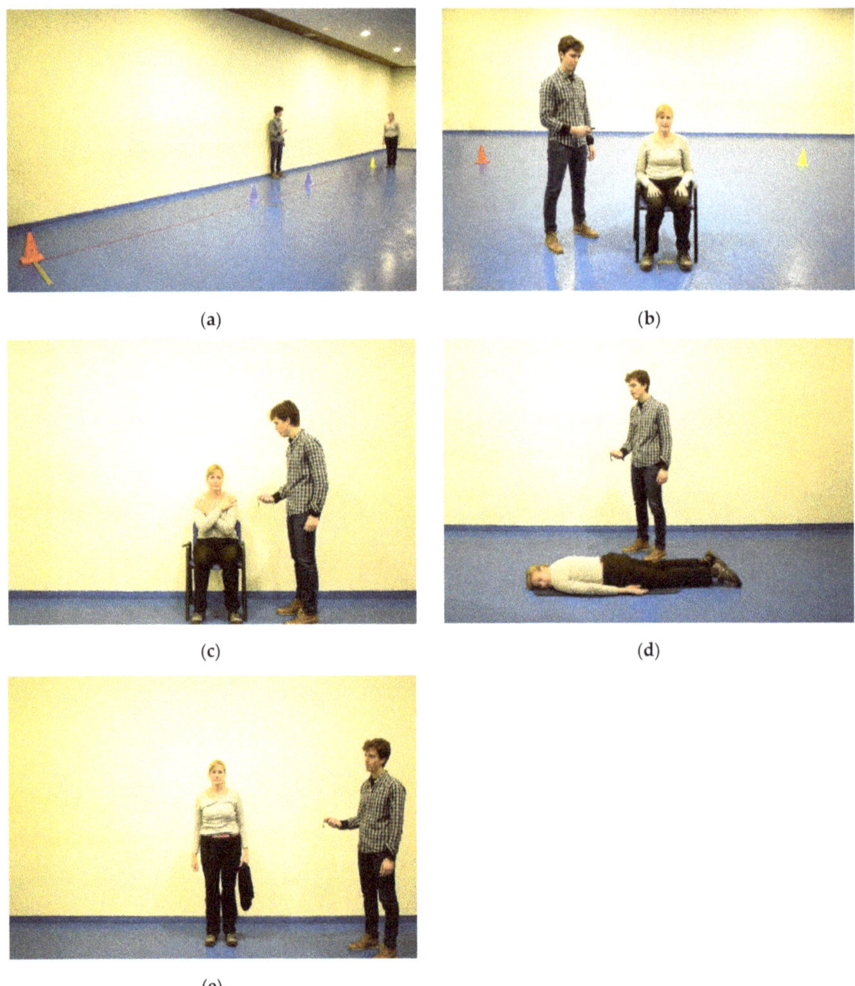

Figure 1. Latin American Group for Maturity (GDLAM) protocol tests in starting position. (**a**) Walk 10 m (W10 m); (**b**) To sit and get up from the chair and move around the house (SCMA); (**c**) Stand up from sitting position (SSP); (**d**) Standing up from the prone position (SPP); (**e**)To put on and take off a T-shirt (PTS).

All tests were performed in two attempts for each individual in a suitable environment, with a minimum interval of five minutes, in which the shortest time in seconds was recorded through a stopwatch. The equipment used consisted of a 48 cm chair (measured from the seat to the floor), a stopwatch (Casio, Malaysia), two cones, a T-shirt, a mat (Olive Fitness, Spain), and a sunny brand metal tape measure. After these tests, the GDLAM index of autonomy (GI) was calculated in scores [16,28], where the lower the value of the score, the better the result, using the following formula. All the tests were measured using time in seconds.

$$GI = [(W10\ m + SSP + SPP + PTS) \times 2] + SCMA]/4$$

Validity and Reliability

The functional autonomy GDLAM protocol underwent a validity analysis using the face validity method [7,17]. A panel composed of three doctors who are specialists in aging and two doctors who are specialists in measures and evaluations, who were not related to the present study to avoid any influence on their opinions, unanimously approved the protocol, with 100% agreement, as a valid instrument to assess the functional autonomy related to the performance of the older people's ADL, applicable to the population of Spain. Before the initial data collection, two experienced evaluators applied the GDLAM protocol tests to 30 older women randomly selected on three different occasions with a minimum interval of 72 h between them to calculate the intraclass correlation coefficient (ICC). The results showed strong reliability with ICC values greater than 0.92 in all tests. One-way ANOVA was used to compare the measurements obtained between the tested occasions and did not show significant differences ($p < 0.05$).

2.3. Statistical Analysis

The data were expressed as mean, standard deviation, and percentage values. The coefficient of variation was applied to analyze the dispersion of the sample data. The normality of the data was verified through the Box-Cox method. The classification levels of the tests and the GI were established by age group every five years and from 50 years of age on, using the percentile (p). It was considered that the lower the value found for the percentile the better the result, according to the classification: Very Good ($p < 0.15$), Good (p 0.16–p 0.50), Regular (p 0.51–p 0.85), and Poor ($p > 0.85$). The one-way analysis of variance (ANOVA) was used, followed by Tukey's post hoc, to identify possible differences in tests and in GI between age groups. Pearson's correlation test was applied for the analysis of associations between age and GDLAM protocol variables. The cut-off threshold for satisfactory GI was determined by the Youden's J statistic (Youden's index) [34,35], this index corresponding values of sensitivity and specificity for groups G1, G2, and G3 (middle-aged women) and G4, G5, G6, and G7 (older women) of the studied population and calculated from the analysis of the receiver operator characteristics (ROC) curve. The study admitted the value of $p < 0.05$ for statistical significance. The data were processed by IBM SPSS Statistics 23 (IBM SPSS, Inc., Chicago, IL, USA).

3. Results

Table 1 presents the anthropometric characteristics of the sample by age categories every 5 years. It is observed that the highest concentration of participants is found in G4 and G5.

Table 1. Characteristics of the sample by age groups.

Age Groups	N	%	Age (Years)	Body Mass (kg)	Height (m)	BMI
G1 (50–54 years)	37	5.22	51.59 ± 1.26	77.18 ± 16.13	1.65 ± 0.10	28.21 ± 4.08
G2 (55–59 years)	36	5.08	56.67 ± 1.35	77.57 ± 12.50	1.65 ± 0.10	28.36 ± 3.40
G3 (60–64 years)	70	9.87	61.69 ± 1.54	76.77 ± 14.64	1.63 ± 0.09	28.72 ± 4.41
G4 (65–69 years)	283	39.92	66.40 ± 1.39	69.93 ± 11.55	1.56 ± 0.08	28.58 ± 4.20
G5 (70–74 years)	160	22.57	71.91 ± 1.45	73.08 ± 11.71	1.57 ± 0.09	29.52 ± 4.08
G6 (75–79 years)	86	12.13	76.09 ± 1.22	73.62 ± 13.29	1.59 ± 0.10	29.13 ± 4.26
G7 (≥80 years)	37	5.22	84.13 ± 3.20	72.57 ± 9.10	1.54 ± 0.08	30.71 ± 4.55
Total	709	100				

BMI: body mass index.

Table 2 presents the performance of the tests and the GI by age categories.

Table 2. Sample results by age groups.

Groups	Tests	Average	Classification
G1 (n = 37; 50–54 years)	W10 m	5.35 ± 1.45	Regular
	SSP	9.38 ± 1.71	Regular
	SPP	2.99 ± 1.22	Regular
	PTS	11.64 ± 2.83	Good
	SCMA	34.72 ± 6.07	Regular
	GI	23.36 ± 3.80	Regular
G2 (n = 36; 55–59 years)	W10 m	5.85 ± 1.73	Regular
	SSP	9.48 ± 2.66	Regular
	SPP	3.02 ± 1.25	Regular
	PTS	12.00 ± 2.28	Regular
	SCMA	36.11 ± 7.04	Regular
	GI	24.20 ± 4.72	Regular
G3 (n = 70; 60–64 years)	W10 m	6.72 ± 1.69	Regular
	SSP	11.24 ± 3.94	Regular
	SPP	4.05 ± 1.91	Regular
	PTS	13.45 ± 3.64	Regular
	SCMA	40.98 ± 7.93	Regular
	GI	27.97 ± 5.82	Regular
G4 (n = 283; 65–69 years)	W10 m	6.95 ± 1.52	Regular
	SSP	11.98 ± 3.19	Regular
	SPP	5.62 ± 3.22	Good
	PTS	15.00 ± 6.12	Regular
	SCMA	42.01 ± 5.78	Regular
	GI	30.28 ± 5.09	Regular
G5 (n = 160; 70–74 years)	W10 m	7.32 ± 1.73	Regular
	SSP	12.17 ± 3.39	Regular
	SPP	6.46 ± 3.56	Regular
	PTS	15.17 ± 5.03	Regular
	SCMA	45.37 ± 9.57	Regular
	GI	31.90 ± 6.42	Regular
G6 (n = 86; 75–79 years)	W10 m	7.88 ± 1.69	Regular
	SSP	12.68 ± 4.63	Regular
	SPP	6.83 ± 3.43	Regular
	PTS	17.44 ± 6.47	Regular
	SCMA	48.63 ± 8.48	Regular
	GI	34.57 ± 6.29	Regular
G7 (n = 37; ≥80 years)	W10 m	8.49 ± 2.70	Regular
	SSP	13.50 ± 3.61	Regular
	SPP	6.90 ± 3.41	Regular
	PTS	20.52 ± 9.08	Regular
	SCMA	56.28 ± 14.49	Regular
	GI	38.77 ± 8.66	Regular

W10 m: walk 10 m; SSP: stand up from sitting position; SPP: standing up from the prone position; PTS: to put on and take off a T-shirt; SCMA: to sit and get up from the chair and move around the house; GI: GDLAM autonomy index.

Some fluctuations of the coefficient of variation (CV%) are observed across the age categories for the different tests, with small increases in SPP and PTS with the age, especially in G7 category (Figure 2).

Figure 2 shows the evolution of the coefficient of variation (CV%) across age categories by GDLAM tests.

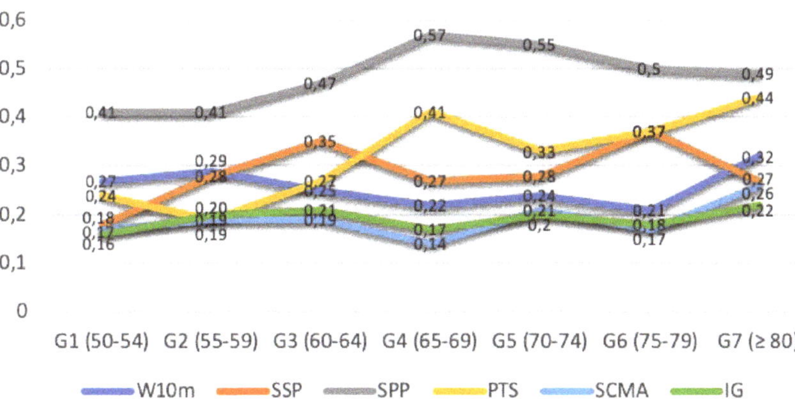

Figure 2. Evolution of the coefficient of variation (CV%) across age categories and test.

Figure 3 shows the distribution of GI scores by age categories and percentile. The GI values raise as the age category increase. It is observed that the tendency curves are similar among all age categories.

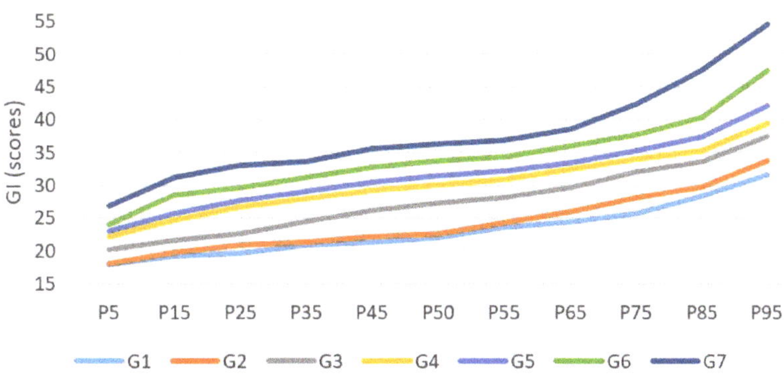

Figure 3. GI Percentile Distribution by Age Categories.

Table 3 presents the comparative results of the tests and the GI of the GDLAM protocol between the age groups. It was observed that the G7 showed longer execution times in the tests W10 m, SSP, SPP, PTS, SCMA, and GI when compared to the groups G1, G2, G3, G4, and G5 ($p < 0.05$). G7 presented lower performance ($p < 0.05$) in PTS, SCMA, and lower GI compared to G6. No significant differences were found between G7 and G6 in the tests W10 m, SSP, and SPP and between G1 and G2 in all tests and in the GI.

Table 3. Comparative analysis of the GDLAM protocol between age groups.

	G1 (50–54 yrs)	G2 (55–59 yrs)	G3 (60–64 yrs)	G4 (65–69 yrs)	G5 (70–74 yrs)	G6 (75–79 yrs)	G7 (≥80 yrs)
W10 m	5.35 ± 1.45 *§†‡	5.85 ± 1.73 *§†‡	6.72 ± 1.69 *§	6.95 ± 1.52 *§	7.32 ± 1.73 *	7.88 ± 1.69	8.49 ± 2.70
SSP	9.38 ± 1.71 *§†‡∂	9.48 ± 2.66 *§†‡∂	11.24 ± 3.94 *§†	11.98 ± 3.19 *	12.17 ± 3.39 *	12.68 ± 4.63	13.50 ± 3.61
SPP	2.99 ± 1.22 *§†‡	3.02 ± 1.25 *§†‡	4.05 ± 1.91 *§†‡	5.62 ± 3.22 *§	6.46 ± 3.56	6.83 ± 3.43	6.90 ± 3.41
PTS	11.64 ± 2.83 *§†‡∂	12.00 ± 2.28 *§†‡	13.45 ± 3.64 *§†‡	15.00 ± 6.12 *§	15.17 ± 5.03 *§	17.44 ± 6.47 *	20.52 ± 9.08
SCMA	34.72 ± 6.07 *§†‡∂	36.11 ± 7.04 *§†‡∂	40.98 ± 7.93 *§†	42.01 ± 5.78 *§†	45.37 ± 9.57 *§	48.63 ± 8.48 *	56.28 ± 14.49
GI	23.36 ± 3.80 *§†‡∂	24.20 ± 4.72 *§†‡∂	27.97 ± 5.82 *§†‡	30.28 ± 5.09 *§†	31.90 ± 6.42 *§	34.57 ± 6.29 *	38.77 ± 8.66

W10 m: walk 10 m; SSP: stand up from sitting position; SPP: standing up from the prone position; PTS: to put on and take off a T-shirt; SCMA: to sit and get up from the chair and move around the house; GI: GDLAM autonomy index. * $p < 0.05$ for G7; § $p < 0.05$ for G6; † $p < 0.05$ for G5; ‡ $p < 0.05$ for G4; ∂ $p < 0.05$ for G3.

Table 4 shows the correlation coefficient between age and functional autonomy, both of the tests and GI. It is observed that all variables of the GDLAM protocol of autonomy presented a positive and significant correlation with age. This shows that the higher the score, the lower the functional autonomy in relation to the tests (in relation to time) and the GI (in relation to the scores). Therefore, in the present study, the higher the age, the longer the time to complete the autonomy tests and the higher the GI score.

Table 4. Correlation analysis between age and functional autonomy (tests and GI).

		W10 m	SSP	SPP	PTS	SCMA	GI
Years	r	0.362	0.227	0.227	0.301	0.481	0.466
	p-value	<0.001	<0.001	<0.001	<0.001	<0.001	<0.001

W10 m: walk 10 m; SSP: stand up from sitting position; SPP: standing up from the prone position; PTS: to put on and take off a T-shirt; SCMA: to sit and get up from the chair and move around the house; GI: GDLAM autonomy index.

Table 5 presents a classification pattern of the GDLAM protocol of functional autonomy for each age group. The classification was made by separate tests and by GI for each age category. It is observed that the lower the values of the tests and the GI, the better the classification.

Table 5. Standardized GDLAM protocol classification of functional autonomy for residents in Spain.

Age Groups		Very Good	Good	Regular	Poor
G1 (n = 37; 50–54 years)	W10 m	<4.17	4.17–4.85	4.86–7.18	>7.18
	SSP	<7.06	7.06–9.01	9.02–11.13	>11.13
	SPP	<1.77	1.77–2.61	2.62–4.85	>4.85
	PTS	<9.95	9.95–11.73	11.74–14.65	>14.65
	SCMA	<29.19	29.19–32.87	32.88–41.41	>41.41
	GI	<19.16	19.16–22.11	22.12–28.35	>28.35
G2 (n = 36; 55–59 years)	W10 m	<4.24	4.24–5.52	5.53–8.00	>8.00
	SSP	<7.60	7.60–9.30	9.31–12.19	>12.19
	SPP	<1.80	1.80–2.87	2.88–4.24	>4.24
	PTS	<8.44	8.44–10.83	10.84–15.45	>15.45
	SCMA	<30.80	30.80–35.26	35.27–45.67	>45.67
	GI	<19.77	19.77–22.68	22.69–29.81	>29.81
G3 (n = 70; 60–64 years)	W10 m	<5.00	5.00–6.65	6.66–8.49	>8.49
	SSP	<8.13	8.13–9.87	9.88–16.36	>16.36
	SPP	<2.16	2.16–3.36	3.37–6.21	>6.21
	PTS	<9.18	9.18–13.19	13.20–17.76	>17.76
	SCMA	<32.45	32.45–39.90	39.91–49.80	>49.80
	GI	<21.66	21.66–27.23	27.24–33.62	>33.62
G4 (n = 283; 65–69 years)	W10 m	<5.40	5.40–6.80	6.81–8.73	>8.73
	SSP	<8.89	8.89–11.61	11.62–15.50	>15.50
	SPP	<3.03	3.03–6.26	6.27–9.58	>9.58
	PTS	<9.97	9.97–14.48	14.49–20.19	>20.19
	SCMA	<36.03	36.03–41.50	41.51–48.03	>48.03
	GI	<24.79	24.79–30.11	30.12–35.40	>35.40
G5 (n = 160; 70–74 years)	W10 m	<5.50	5.50–7.12	7.13–8.91	>8.91
	SSP	<8.92	8.92–11.91	11.92–15.04	>15.04
	SPP	<3.63	3.63–4.91	4.92–7.90	>7.90
	PTS	<10.49	10.49–13.79	13.80–20.08	>20.08
	SCMA	<37.22	37.22–44.43	44.44–51.35	>51.35
	GI	<25.73	25.73–31.49	31.50–37.50	>37.50
G6 (n = 86; 75–79 years)	W10 m	<6.14	6.14–7.45	7.46–9.45	>9.45
	SSP	<9.15	9.15–12.13	12.14–16.90	>16.90
	SPP	<3.85	3.85–5.36	5.37–9.60	>9.60
	PTS	<11.87	11.87–15.30	15.31–24.22	>24.22
	SCMA	<41.52	41–52–47.18	47.19–57.89	>57.89
	GI	<28.50	28.50–33.83	33.84–40.46	>40.46
G7 (n = 37; ≥80 years)	W10 m	<6.12	6.12–7.94	7.95–11.16	>11.16
	SSP	<9.87	9.87–13.48	13.49–16.33	>16.33
	SPP	<4.41	4.41–5.90	5.91–10.67	>10.67
	PTS	<12.43	12.43–18.47	18.48–29.53	>29.53
	SCMA	<42.46	42.46–52.67	52.68–75.04	>75.04
	GI	<31.28	31.28–36.30	36.31–47.65	>47.65

W10 m: walk 10 m; SSP: stand up from sitting position; SPP: standing up from the prone position; PTS: to put on and take off a T-shirt; SCMA: to sit and get up from the chair and move around the house; GI: GDLAM autonomy index. Tests: values are in seconds; GI: values are units of the scores.

Figure 4A,B shows the analysis of the ROC curve for the prediction of middle-aged (G1, G2, and G3: 50–64 years) and older individuals (G4, G5, G6, and G7: ≥65 years) with low functional autonomy. The GI cutoff points >28 and >34 showed strong sensitivity (99.8) and specificity (89.6) for the group of middle age and strong sensitivity (99.7) and specificity (87.6) for the group of older women.

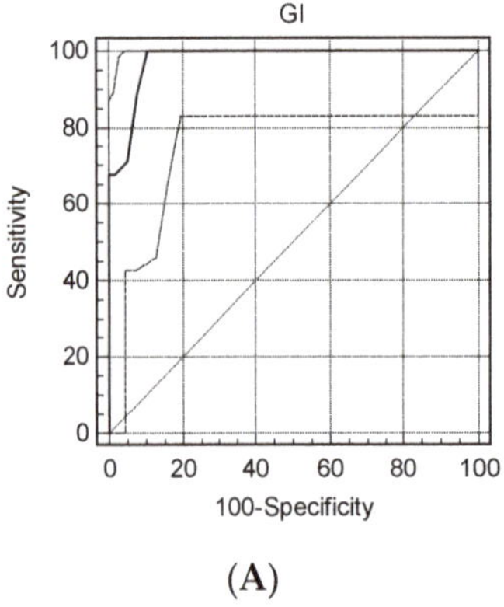
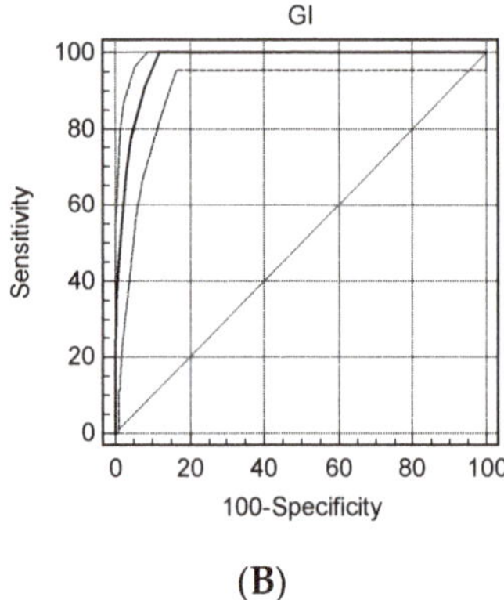

(A) **(B)**

Figure 4. (**A**,**B**). Analysis of the ROC curve with GI cutoff point for prediction of middle-aged (A: >28; Sensitivity = 99.8; specificity = 89.6; area = 0.98; CI (95%) = 0.93–0.99; $p < 0.001$.) and older individuals with low functional autonomy (B: >34; Sensitivity = 99.7; specificity = 87.6; area = 0.97; CI (95%) = 0.95–0.98; $p < 0.001$).

4. Discussion

The GDLAM protocol was applied to assess ADL performance in middle-aged and older women in Spain. The tests of this protocol are similar to ADL, such as crossing a street, sitting and getting up from a chair, walking short distances around the house and with changes of direction, dressing, among other activities. This similarity is important since functional autonomy is associated with ADL performance. The components of physical fitness of this protocol involve speed, agility, balance, coordination, power, and resistance of lower and upper limbs in ADL [7,17].

Some previous studies have used the GDLAM protocol to assess functional autonomy levels in elderly patients with chronic obstructive pulmonary disease [18], postmenopausal women [19], postmenopausal women with Parkinson's disease [21], and apparently healthy elderly women [20,22–25], and to analysis the effect of resistance training [19,20,22,25], Pilates [20], aquatic [21,23], and walking exercises [24] interventions. The standard times and the scores provided by the GDLAM protocol to each age group allows the evaluation of Spanish middle-aged and older women. Consequently, the standard values of the present study contribute with accurate standards to categorize functional autonomy and could also be helpful to check the efficiency of exercise training programs [17].

Walking 10 m is important for the individual to have the security and independence to cross a street or move from one room to another in the house. Araújo-Gomes et al. [36] applied resistance exercises (3 days per week) and Pilates (2 days per week) in 12 postmenopausal women (age = 59.83 ± 5.0 years). After 16 weeks of intervention, those women improved the performance in this task (time: 5.71 s). According to the standardized GDLAM protocol classification of the present study, this time is considered "Regular" (5.53–8.00) for the age range of 55 to 59 years and "Good" (5.00–6.65) for the age range of 60 to 64 years. Similarly, in the standardized GDLAM protocol classification of Dantas et al. [15] the age range of 60 to 64 years would also be classified as "Good" (5.52–7.04). However, this classification [15] only included elderly Brazilian women (≥60 years).

Sitting and getting up as well as getting up and moving around the house are movements involved in everyday tasks. For instance, sitting and getting up from a chair or a sofa and going from place to place. Performing these activities satisfactorily requires adequate levels of muscle strength and power, mainly in the lower limbs, in addition to balance and agility. Likewise, the capacity of standing up from the prone position also requires these physical capabilities. The SCMA, SSP, and SPP tests of the GDLAM protocol are used to analyze these variables [17]. Vale et al. [24] found significant decreases in the time to execute these tasks (SCMA: 44.17 ± 3.08 s; SSP: 9.84 ± 1.56; SPP: 3.01 ± 0.56 s) in a group of 15 elderly women (age: 68 ± 4.4 years) who practiced 24 weeks of resistance exercises (2×/week; 50 min/session).

Putting on and taking off a T-shirt is a daily process throughout life. Sometimes, middle-aged and elderly women present difficulty performing this movement, as they may have some osteoarticular limitation or some functional decrease. Resistance [19,20,22,25] and aquatic exercises [21,23] can improve strength and flexibility. Thus, they can influence the performance of this daily activity, improving functional autonomy. A randomized controlled trial [20] found significant decreases in the time to perform the PTS test in older women from Murcia, Spain, after 36 weeks of interventions (2×/week; 60 min/session). Those positive outcomes were observed both in the Pilates (n = 20; age: 67.5 ± 3.87 years; PTS: 15.53 ± 2.98 s) and the resistance exercises (n = 20; age: 73.36 ± 4.84 years; PTS: 16.24 ± 3.82 s) groups. As reported by the standardized GDLAM protocol classification of the current study, the Pilates and the resistance group are classified as "Regular" (65–69 years: 14.49–20.19 s; 70–74 years: 13.80–20.08 s) in this test.

The five tests of the protocol establish a general indicator of functional autonomy, the GI. The results obtained in the tests of the GDLAM protocol showed strong reproducibility, which increases the precision and accuracy of the data to establish the adequate standardized classification of functional autonomy through the GI. Tests and GI outcomes of the present study showed that, as age increases, the time to perform ADL also increases. This occurs due to the deleterious effects of aging and the lifestyle adopted by individuals, comprising the sedentary behavior [37].

The GI cutoff point provided for prediction of low autonomy can be very useful for healthcare and educational settings. Women with higher scores of the cutoff points should be encouraged to change their lifestyle to prevent dependency in later life. In middle-aged women, GI values above 28 (scores) may indicate the need for specific care and changes in lifestyle to avoid a sedentary lifestyle and aging with low functional autonomy. For older women, GI values above 34 (scores), probably show the necessity for regular physical exercises to maintain ADL. Changes in posture, as one of the aging characteristics, especially from the age of 80, can increase the time of execution of the ADL since, in the present study, an increase in total body mass with lower height was observed in G7. This may be related to the loss of bone mineral density due to aging and intervertebral discs degeneration that causes an increase in spinal curvatures, such as kyphosis intensification [38].

Another important issue in this context is the components of physical fitness that affect health and physical function, such as cardiorespiratory fitness, muscle strength and endurance (muscular fitness), flexibility, neuromotor fitness, and body composition [4]. Therefore, the American College of Sports Medicine (ACSM) [3] and the World Health Organization (WHO) [39] highlight that older people should remain physically active to preserve or even improve skeletal integrity. Ideally, exercise programs for older people should embrace not only weight-bearing endurance and resistance activities to preserve bone mass and to counteract muscle strength decline and muscle mass loss, but also activities intended to increase balance and prevent falls [40–42], which involves ADL performance and functional autonomy.

An active lifestyle can promote healthy aging and greater independence in old age [3,43]. This way, the percentile values adopted to establish the classification of the battery of tests and the GI of the GDLAM protocol can allow the monitoring of individuals to maintain an active lifestyle.

The strengths of the present study are that the reference values of the GDLAM functional autonomy assessment battery did not exist in the Spanish context and the inclusion of a large population sample size, although it is not a representative sample of the Spanish population. As limitations of the study, it must be indicated that it was only conducted with middle-aged and older women. A future line of research is to replicate the study with men and verify if there are differences by sex. Another limitation is that the number of participants being lower in some of the age groups, due to a lack of interest in participation, which may result in less accurate policy data. This suggests the need to continue researching the motives for practice and motivation of these groups, in order to offer them exercise programs that will achieve their adherence.

5. Conclusions

This is the first study providing reference values (or percentiles) for the GDLAM protocol in a large cohort of middle-aged and older Spanish women. Age-specific functional autonomy normative values for middle-aged and older Spanish women have been established. The normative values hereby provided will enable evaluation and adequate interpretation of middle-aged and older Spanish women. This "tool" is especially interesting in healthcare and educational settings for healthy and active aging. Normative values can help clinicians and physical education trainers to know the level of functional autonomy of women practitioners and prescribe exercise programs that fit their level and state of health, to help them improve. In addition, the reported normative values should be used to motivate practitioners to do regular physical activity and increase tFheir functional autonomy for better health and quality of life.

Author Contributions: Conceptualization, P.J.M.-P. and R.G.d.S.V.; methodology, P.J.M.-P., N.G.-G., D.J.-P., J.B.P.d.C. and R.G.d.S.V.; software, J.B.P.d.C., R.G.d.S.V.; validation, P.J.M.-P., D.J.-P., J.B.P.d.C. and R.G.d.S.V.; formal analysis, P.J.M.-P., J.B.P.d.C. and R.G.d.S.V.; investigation, P.J.M.-P., N.G.-G., R.V.-C., G.M.G.-G., A.L.-V., A.E.-G., D.J.-P., A.C.-B., D.V.-D., J.B.P.d.C. and R.G.d.S.V.; resources, P.J.M.-P., N.G.-G., R.V.-C., G.M.G.-G., A.L.-V., A.E.-G., D.J.-P., A.C.-B., D.V.-D., J.B.P.d.C. and R.G.d.S.V.; data curation, P.J.M.-P. and R.G.d.S.V.; writing—original draft preparation, P.J.M.-P., N.G.-G., R.V.-C., D.J.-P., J.B.P.d.C. and R.G.d.S.V.; writing—review and editing, P.J.M.-P., N.G.-G., R.V.-C., G.M.G.-G., A.L.-V., A.E.-G., D.J.-P., A.C.-B., D.V.-D., J.B.P.d.C. and R.G.d.S.V.; visualization, P.J.M.-P., N.G.-G., R.V.-C., D.J.-P., D.V.-D., J.B.P.d.C. and R.G.d.S.V.; supervision, P.J.M.-P., N.G.-G., D.J.-P., and R.G.d.S.V.; project administration, P.J.M.-P.; funding acquisition, P.J.M.-P. All authors have read and agreed to the published version of the manuscript.

Funding: The present research on active aging of members of HEALTHY-AGE Network (reference 08/UPR/20) is supported by a grant from the Spanish Ministry of Culture and Sport- Sports Sciences Networks and PJMP was supported by a grant from the European Union's ERASMUS + SPORT program (reference: 603121-EPP-1-2018-1-ES-SPO-SCP). DJP was supported by a grant from the Spanish Ministry of Science and Innovation - MINECO (RYC-2014-16938).

Acknowledgments: The research team would like to thank the heads of the social and women's centers and all the older women for their participation in this research, and Catholic University San Antonio of Murcia (UCAM) for its support to the line of research on healthy and active ageing.

Conflicts of Interest: The authors declare no conflict of interest.

References

1. El Khoudary, S.R.; Greendale, G.; Crawford, S.L.; Avis, N.E.; Brooks, M.M.; Thurston, R.C.; Karvonen-Gutierrez, C.; Waetjen, L.E.; Matthews, K. The menopause transition and women's health at midlife. *Menopause* **2019**, *26*, 1213–1227. [CrossRef]

2. Daniel, F.D.N.R.; Vale, R.G.D.S.; Giani, T.S.; Bacellar, S.; Escobar, T.; Stoutenberg, M.; Dantas, E.H.M. Correlation between static balance and functional autonomy in elderly women. *Arch. Gerontol. Geriatr.* **2011**, *52*, 111–114. [CrossRef]

3. Chodzko-Zajko, W.J.; Proctor, D.N.; Singh, M.A.F.; Minson, C.T.; Nigg, C.R.; Salem, G.J.; Skinner, J.S. Exercise and Physical Activity for Older Adults. *Med. Sci. Sports Exerc.* **2009**, *41*, 1510–1530. [CrossRef]

4. Garber, C.E.; Blissmer, B.; Deschenes, M.R.; Franklin, B.A.; Lamonte, M.J.; Lee, I.M.; Nieman, D.C.; Swain, D.P. American College of Sports Medicine. American College of Sports Medicine position stand. Quantity and quality of exercise for developing and maintaining cardiorespiratory, musculoskeletal, and neuromotor fitness in apparently healthy adults: Guidance for prescribing exercise. *Med. Sci. Sports Exerc.* **2011**, *43*, 1334–1359. [CrossRef] [PubMed]

5. Lee, I.-M.; Shiroma, E.J.; Lobelo, F.; Puska, P.; Blair, S.N.; Katzmarzyk, P.T. Effect of physical inactivity on major non-communicable diseases worldwide: An analysis of burden of disease and life expectancy. *Lancet* **2012**, *380*, 219–229. [CrossRef]

6. Perrig-Chiello, P.; Perrig, W.J.; Uebelbacher, A.; Stähelin, H.B. Impact of physical and psychological resources on functional autonomy in old age. *Psychol. Health Med.* **2006**, *11*, 470–482. [CrossRef]

7. Dantas, E.H.M.; Vale, R.G.D.S. GDLAM'S protocol of functional autonomy evaluation. *Fit. Perform. J.* **2004**, *3*, 175–183. [CrossRef]

8. Fernández-García, Á.; Gómez-Cabello, A.; Moradell, A.; Navarrete-Villanueva, D.; Pérez-Gómez, J.; Ara, I.; Pedrero-Chamizo, R.; Subías-Perié, J.; Muniz-Pardos, B.; Casajus, J.; et al. How to Improve the Functional Capacity of Frail and Pre-Frail Elderly People? Health, Nutritional Status and Exercise Intervention. The EXERNET-Elder 3.0 Project. *Sustainability* **2020**, *12*, 6246. [CrossRef]

9. Kojima, G. Frailty as a predictor of disabilities among community-dwelling older people: A systematic review and meta-analysis. *Disabil. Rehabilit.* **2016**, *39*, 1897–1908. [CrossRef] [PubMed]

10. Clegg, A.; Young, J.; Iliffe, S.; Rikkert, M.O.; Rockwood, K. Frailty in elderly people. *Lancet* **2013**, *381*, 752–762. [CrossRef]

11. Morley, J.E.; Vellas, B.; Van Kan, G.A.; Anker, S.D.; Bauer, J.M.; Bernabei, R.; Cesari, M.; Chumlea, W.; Doehner, W.; Evans, J.; et al. Frailty Consensus: A Call to Action. *J. Am. Med. Dir. Assoc.* **2013**, *14*, 392–397. [CrossRef]

12. Da Silva, G.C.; Neto, J.L.C. Quality of life and functional capability of elderly Brazilian women. *Work* **2019**, *62*, 97–106. [CrossRef] [PubMed]

13. Pérez-Díaz, J.; Abellán-García, A.; Aceituno-Nieto, P.; Ramiro-Fariñas, D. Un Perfil de las Personas Mayores en España, 2020. Indicadores Estadísticos Básicos. Madrid, Informes Envejecimiento en Red n° 25, 38p. Available online: http://envejecimientoenred.es/un-perfil-de-las-personas-mayores-en-espana-2020-indicadores-estadisticos-basicos/ (accessed on 18 August 2020).

14. Mayo, X.; Luque-Casado, A.; Jimenez, A.; Del Villar, F. Physical Activity Levels for Girls and Young Adult Women versus Boys and Young Adult Men in Spain: A Gender Gap Analysis. *Sustainability* **2020**, *12*, 6265. [CrossRef]

15. Cunningham, C.; Sullivan, R.O.; Caserotti, P.; Tully, M.A. Consequences of physical inactivity in older adults: A systematic review of reviews and meta-analyses. *Scand. J. Med. Sci. Sports* **2020**, *30*, 816–827. [CrossRef]

16. Alcañiz, M.; Brugulat, P.; Guillén, M.; Medina-Bustos, A.; Mompart-Penina, A.; Solé-Auró, A. Risk of dependence associated with health, social support, and lifestyle. *Rev. Saúde Públ.* **2015**, *49*, 26. [CrossRef] [PubMed]

17. Dantas, E.H.M.; Figueira, H.A.; Emygdio, R.F.; Vale, R.G.S. Functional Autonomy GdlAm Protocol Classification Pattern in Elderly Women. *Indian J. Appl. Res.* **2011**, *4*, 262–266. [CrossRef]

18. Boechat, F.; Vale, R.G.D.S.; Dantas, E.H. Evaluación de la autonomía funcional de ancianos con EPOC mediante el protocolo GDLAM. *Rev. Esp. Geriatr. Gerontol.* **2007**, *42*, 251–253. [CrossRef]

19. Borba-Pinheiro, C.J.; Dantas, E.H.M.; Vale, R.G.D.S.; Drigo, A.J.; Carvalho, M.C.G.D.A.; Tonini, T.; Meza, E.I.A.; De Figueiredo, N.M.A. Resistance training programs on bone related variables and functional independence of postmenopausal women in pharmacological treatment: A randomized controlled trial. *Arch. Gerontol. Geriatr.* **2016**, *65*, 36–44. [CrossRef]

20. Carrasco-Poyatos, M.; Rubio-Arias, J.Á.; Ballesta-García, I.; Ramos-Campo, D.J. Pilates vs. muscular training in older women. Effects in functional factors and the cognitive interaction: A randomized controlled trial. *Physiol. Behav.* **2019**, *201*, 157–164. [CrossRef]

21. Hall-López, J.A.; Ochoa-Martínez, P.Y.; Monzon, C.O.L.; Vale, R.G.S. Effects of four months of periodized aquatic exercise program on functional autonomy in post-menopausal women with Parkinson's disease. *Retos* **2018**, *33*, 217–220.

22. Pereira, F.F.; Monteiro, N.; Vale, R.G.D.S.; Gomes, A.L.M.; Novaes, J.D.S.; Júnior, A.G.D.F.; Dantas, E.H.M. Efecto del entrenamiento de fuerza sobre la autonomía funcional en mujeres mayores sanas. *Rev. Esp. Geriatr. Gerontol.* **2007**, *42*, 342–347. [CrossRef]

23. Pernambuco, C.S.; Borba-Pinheiro, C.J.; Vale, R.G.D.S.; Di Masi, F.; Monteiro, P.K.P.; Dantas, E.H. Functional autonomy, bone mineral density (BMD) and serum osteocalcin levels in older female participants of an aquatic exercise program (AAG). *Arch. Gerontol. Geriatr.* **2013**, *56*, 466–471. [CrossRef] [PubMed]

24. Vale, R.G.S.; Castro, J.B.P.; Mattos, R.S.; Rodrigues, V.F.; Oliveira, F.B.; Rosa, G.; Gama, D.; Nunes, R.A.; Boechat, F. Analysis of balance, muscle strength, functional autonomy, and quality of life in elderly women submitted to a strength and walking program. *J. Exerc. Physiol.* **2018**, *21*, 13–24.

25. Marcos-Pardo, P.J.; Orquin-Castrillón, F.J.; Gea-García, G.M.; Menayo-Antúnez, R.; González-Gálvez, N.; Vale, R.G.D.S.; Martínez-Rodríguez, A. Effects of a moderate-to-high intensity resistance circuit training on fat mass, functional capacity, muscular strength, and quality of life in elderly: A randomized controlled trial. *Sci. Rep.* **2019**, *9*, 1–12. [CrossRef] [PubMed]

26. World Medical Association (WMA). Declaration of Helsinki. Ethical Principles for Medical Research Involving Human Subjects. *Jahrb. Wiss. Ethik* **2009**, *14*. [CrossRef]

27. Marfell-Jones, M.; Stewart, A.D.; Ridder, J.H. *International Standards for Anthropometric Assessment*; International Society for the Advancement of Kinanthropometry: Wellington, New Zealand, 2012.

28. Vale, R.G.S. Avaliação da autonomia funcional do idoso. *Fit. Perf. J.* **2005**, *4*, 4.

29. Sipilä, S.; Multanen, J.; Kallinen, M.; Era, P.; Suominen, H. Effects of strength and endurance training on isometric muscle strength and walking speed in elderly women. *Acta Physiol. Scand.* **1996**, *156*, 457–464. [CrossRef]

30. Andreotti, R.A.; Okuma, S.S. Validating a test battery of activities of daily living for physically independent elderly. *Rev. Paul. Educ. Física* **1999**, *13*, 46. [CrossRef]

31. Guralnik, J.M.; Simonsick, E.M.; Ferrucci, L.; Glynn, R.J.; Berkman, L.F.; Blazer, D.G.; Scherr, P.A.; Wallace, R.B. A Short Physical Performance Battery Assessing Lower Extremity Function: Association with Self-Reported Disability and Prediction of Mortality and Nursing Home Admission. *J. Gerontol.* **1994**, *49*, M85–M94. [CrossRef]

32. Alexander, N.B.; Ulbrich, J.; Raheja, A.; Channer, D. Rising from the floor in older adults. *J. Am. Geriatr. Soc.* **1997**, *45*, 564–569. [CrossRef]

33. Vale, R.G.S.; Pernambuco, C.S.; Novaes, J.S.; Dantas, E.H.M. Functional autonomy test: To dress and undress a sleeveless shirt (DUSS). *Rev. Bras. Ciênc. Mov.* **2006**, *14*, 71–78.

34. Youden, W.J. Index for rating diagnostic tests. *Cancer* **1950**, *3*, 32–35. [CrossRef]

35. Borges, L.S.R. Diagnostic Accuracy Measures in Cardiovascular Research. *Int. J. Cardiovasc. Sci.* **2016**, *29*, 218–222. [CrossRef]

36. Araújo-Gomes, R.C.; Valente-Santos, M.; Vale, R.G.D.S.; Drigo, A.J.; Borba-Pinheiro, C.J. Effects of resistance training, tai chi chuan and mat pilates on multiple health variables in postmenopausal women. *J. Hum. Sport Exerc.* **2019**, *14*, 122–139. [CrossRef]

37. Wullems, J.A.; Verschueren, S.M.P.; Degens, H.; Morse, C.I.; Onambélé, G.L. A review of the assessment and prevalence of sedentarism in older adults, its physiology/health impact and non-exercise mobility counter-measures. *Biogerontology* **2016**, *17*, 547–565. [CrossRef]

38. Vo, N.V.; Hartman, R.A.; Patil, P.R.; Risbud, M.V.; Kletsas, D.; Iatridis, J.C.; Hoyland, J.A.; Le Maitre, C.L.; Sowa, G.A.; Kang, J.D. Molecular mechanisms of biological aging in intervertebral discs. *J. Orthop. Res.* **2016**, *34*, 1289–1306. [CrossRef]

39. Nawrocka, A.; Mynarski, W. Objective Assessment of Adherence to Global Recommendations on Physical Activity for Health in Relation to Spirometric Values in Nonsmoker Women Aged 60–75 Years. *J. Aging Phys. Act.* **2017**, *25*, 123–127. [CrossRef]

40. Kohrt, W.M.; Bloomfield, S.A.; Little, K.D.; Nelson, M.E.; Yingling, V.R. Physical Activity and Bone Health. *Med. Sci. Sports Exerc.* **2004**, *36*, 1985–1996. [CrossRef]

41. Fragala, M.S.; Cadore, E.L.; Dorgo, S.; Izquierdo, M.; Kraemer, W.J.; Peterson, M.D.; Ryan, E.D. Resistance Training for Older Adults. *J. Strength Cond. Res.* **2019**, *33*, 2019–2052. [CrossRef]

42. Battaglia, G.; Giustino, V.; Messina, G.; Faraone, M.; Brusa, J.; Bordonali, A.; Barbagallo, M.; Palma, A.; Dominguez, L.-J. Walking in Natural Environments as Geriatrician's Recommendation for Fall Prevention: Preliminary Outcomes from the "Passiata Day" Model. *Sustainability* **2020**, *12*, 2684. [CrossRef]

43. Roberts, C.E.; Phillips, L.H.; Cooper, C.L.; Gray, S.; Allan, J.L. Effect of Different Types of Physical Activity on Activities of Daily Living in Older Adults: Systematic Review and Meta-Analysis. *J. Aging Phys. Act.* **2017**, *25*, 653–670. [CrossRef] [PubMed]

Publisher's Note: MDPI stays neutral with regard to jurisdictional claims in published maps and institutional affiliations.

Article

Validity of the Portable Ultrasound BodyMetrix™ BX-2000 for Measuring Body Fat Percentage

Seoungki Kang [1], Jeong-Hui Park [2], Myong-Won Seo [3], Hyun Chul Jung [4], Yong Ik Kim [5] and Jung-Min Lee [2,*]

[1]   Graduate School of Education, Yongin University, 134 Yongindaehak-ro, Cheoin-gu, Yongin-si 17092, Gyeonggi-do, Korea; ksk0527@hanmail.net

[2]   Department of Physical Education, Kyung Hee University (Global Campus), 1732 Deokyoungdaero, Giheung-gu, Yongin-si 17014, Gyeonggi-do, Korea; jeonghee@khu.ac.kr

[3]   Department of Taekwondo, Kyung Hee University (Global Campus), 1732 Deokyoungdaero, Giheung-gu, Yongin-si 17014, Gyeonggi-do, Korea; myongwonseo@khu.ac.kr

[4]   Department of Coaching, Kyung Hee University (Global Campus), 1732 Deokyoungdaero, Giheung-gu, Yongin-si 17014, Gyeonggi-do, Korea; jhc@khu.ac.kr

[5]   Department of Sports Health and Rehabilitation, Kookmin University, 77 Jeongneung-Ri, Seongbuk-gu, Seoul 02707, Korea; hyyongik@naver.com

*   Correspondence: jungminlee@khu.ac.kr; Tel.: +82-31-201-2736

Received: 8 September 2020; Accepted: 20 October 2020; Published: 22 October 2020

Abstract: BodyMetrix™ BX-2000 (IntelaMetrix, Livermore, CA, USA) has been introduced as one of the alternatives and portable methods to estimate body fat percentage. However, inconsistent results between protocols built-in the Bodymetrix™ may be compelling the question of its validity. Thus, this study first investigated the possible errors between protocols and evaluated the validity of body fat percentage (BF%) compared to the gold standard method (dual-energy X-ray absorptiometry, DEXA). One hundred and five collegiate males, aged 20.01 ± 2.11 years, body height, 174.81 ± 6.01 cm, body mass, 73.26 ± 13.60 kg, and body mass index, 23.91 ± 3.77 kg·m^{-2} participated in the present study. Participants' body fat percentage was estimated by built-in nine different protocols in the BodyMetrix™ BX-2000 using A-MODE ultrasound. Pearson correlation (r), Mean absolute percentage errors (MAPEs), Bland & Altman plots, and Equivalence testing were used to examine the validity of each protocol by comparing it to the criterion measure (i.e., DEXA). The results indicated good potential for almost all of the protocols in correlation (Min: r = 0.79, Max: r = 0.92)., MAPEs (Min: 20.0%, Max: 33.8%), and Bland-Altman (Min diff: 16.7, Max diff: 41.4). Particularly, the estimated BF% from protocol 7 (4-sites by Durnin & Wormersley) and protocol 9 (9-sites Parllo) were completed within the equivalence zone (±10% of the mean). The estimates measured by protocol 7 and protocol 9 identified as the most valid methods for estimating BF% using a BodyMetrix™ BX-2000, compared to the DEXA. Our findings provide valuable information when applying in young male individuals, but future studies with other populations such as female or adolescents may be required to suggest a valid protocol within the instrument.

Keywords: body composition; body fat percentage; DEXA; body metrix

---

1. Introduction

The World Health Organization referred to obesity as a "New Infectious Disease of the 21st century" [1], however, complex social structures and dietary changes affect people' lifestyle, resulting in a steady increase in obesity due to lack of exercise, malnutrition, and stress [2]. Also, obesity is associated with insulin resistance that increases cardiovascular diseases [3], and it makes us suffer from diseases such as metabolic syndrome, coronary artery disease, type 2 diabetes, hyperlipidemia, and cancer [4].

With increasing age, individuals are more easily exposed to the changes that adversely affect individuals' health and are at an increased risk of gaining weight [5]. In particular, obesity in early adulthood is likely to lead to obesity in adulthood, forming both healthy body weight and healthy body composition is very important [5]. One of the strategies that are promising in preventing obesity is to pinpoint critical periods of weight gain. One such recognized critical period of weight gain is the student transition from high school to college life [6]. Also, Gropper and colleagues said that it is important for young adults to maintain their health with proper weight status because adults aged 18 and 25 show a significant weight gain due to changes in the amount of stored body fat to their weight [7]. Unfortunately, once individuals start gaining weight, it becomes hard to reverse and increases vulnerability to obesity.

Most college student access to obtain information on their body composition using Body Mass Index (BMI) classification charts, however, BMI has demonstrated that it is not a reliable method because it does not clearly classify personal characteristics and it could not accurately measure adipose tissue percentage [8–10]. Therefore, it has been recently reported that Dual-Energy X-ray Absorptiometry (DEXA) and BodyMetrix™ BX-2000 (IntelaMetrix, Livermore, CA, USA) are the best way to measure body fat as an alternative method [10,11]. DEXA is a standard method which has high accuracy and reliability with lower error than 2% in measuring body fat percentage (BF%), and the measurements using ultrasound like BodyMetrix™ BX-2000 also is highly evaluated for its utilization because it is portable, easy to use, inexpensive, and accurate in measuring BF%. According to the Smith and his colleagues' study, they demonstrated that BodyMetrix™ BX-2000 ultrasound device was a valid and reliable standard testing tool for measuring BF% [12]. In addition, there are some recent studies for comparing DEXA and each of the 9 protocols in BodyMetrix™ BX-2000. Lyon and colleagues verified the validity of the 7-sites protocol by Jackson & Pollock using BodyMetrix™ BX-2000 compared to the DEXA [13]. Also, Ulbricht et al. indicated that the 9-sites Parllo protocol and DEXA had high validity [14], and Kang and his colleagues' study also verified that there was high validity between 3-sites protocol by Pollock of BodyMetrix™ BX2000 and DEXA [15].

However, there were inconsistencies among the available studies on comparing the different protocols using BodyMetrix™ BX-2000. Johnson and his colleagues evaluated between 7-sites by Jackson & Pollock protocol in BodyMetrix™ BX-2000 compared to the DEXA, and they indicated that Jackson & Pollock protocol in BodyMetrix™ BX-2000 were underestimated for the BF% [16]. Loenneke et al. found that 1-point Bicep protocol and 3-sites protocol by Jackson & Pollock in BodyMetrix™ BX-2000 significantly underestimated BF% when compared with DEXA [17]. Therefore, the purpose of the present study systemically investigates possible errors in different protocols and to demonstrate the validity of BF% between each different protocol in BodyMetrix™ BX2000 compared to the criterion method (i.e., DEXA) through the novel statistical analysis (i.e, Equivalence test).

2. Materials and Methods

2.1. Participants

One hundred-five college male students participated in the present study. All participants provided their written informed consent after obtaining a full explanation of the study's purpose, benefit, risk, and procedure. The present study was approved by the Institutional Review Board of Yongin University (2-1040966-AB-N-01-20-1810-HSR-113-8).

2.2. Anthropometric and Body Composition

The height and body weight were assessed using a stadiomet and a digital scale, respectively. The body mass index (BMI) was calculated as weight (kilogram) divided by the height in square meters. Body fat percentage was measured with a dual-energy X-ray absorptiometry (DEXA; GE Healthcare, Madison, WI, USA). DEXA was performed in barefoot and to remove the metal substance from wearing light clothing and their body for whole-body scan area.

2.3. BodyMetrix™ BX2000

The present study utilized BodyMetrix™ BX2000 as a tool using ultrasound waves (IntelaMetrix, Brentwood, CA, USA). This ultrasound device used the principle of reflection that some waves penetrated, and others reflected the probe when sonic waves move into the border of the two mediums with different acoustic resistances. A-MODE was used by ultrasound imaging method and it is called by amplitude mode compared to the B-MODE which only displays the brightness mode. A-MODE is the most basic form among ultrasound imaging methods that indicate the intensity of the reflected sound in time (distance) and is also an effective method for measuring distances. It transmits high-frequency sound waves to penetrate. Through body tissues with A-Mode 2.5 Mhz portable ultrasound. The differentiation of the body tissue interface is determined based on the return times of ultrasound to reflect though the sound head transducer. In addition, the primary advantage of this device is that it can minimize human measurement errors, unlike the skin fold measure. There are 11 protocols for ultrasound measurements in males, however, the present study used only 9 protocols except for two protocols developed for adolescents (2-site by slaughter 8–17 (yrs), 2-sites by slaughter for children). Table 1 shows 9 protocols and anatomical sites.

Table 1. Measuring sites of each protocol.

BodyMetrix™ BX-2000		
No.	**Protocols**	**Measuring Sites**
1	1-point Bicep	Bicep
2	2-sites by A. W. Sloan	Thigh, Scapula
3	3-sites by Jackson & Pollock	Thigh, Chest, Waist
4	3-sites by Pollock	Chest, Scapula, Tricep
5	4-sites NHCA Formula	Chest, Scapula, Axilla, Waist
6	4-sites by Forsyth-Sinning	Waist, Tricep, Scapula, Axilla
7	4-sites by Durnin & Wormersley	Bicep, Scapula, Tricep, Hip
8	7-sites by Jackson & Pollock	Chest, Scapula, Axilla, Tricep, Waist, Hip, Thigh
9	9-site Parllo	Chest, Tricep, Bicep, Scapula, Lower Back, Hip, Waist, Thigh, Calf

2.4. Data Analysis

Demographic information (i.e., Age) and participants' body composition (i.e., Height, Weight, and BMI) were examined by descriptive statistics using SPSS 25.0 version. Pearson correlations were used to investigate the association between each protocol and the criterion method (i.e., DEXA). Mean absolute percent errors (MAPE) were calculated to provide the overall measurement error of 9 protocols. The mean absolute percentage errors are widely used performance evaluation indices in prediction. This is a more conservative estimate of error that takes into account both over-and under-estimation because of using the absolute value in the calculation. Repeated measure ANOVA to determine if there were any statistical differences between these 9 protocols and inter-individual differences were controlled for. Furthermore, the novel statistical approach, 'equivalence testing', was utilized in this study to examine measurement agreements between the protocols and the DEXA. Unlike traditional testing, the estimate was considered equivalent if the 90% confidence interval (CI) for the mean was included in the proposed equivalence zone (e.g., ±10% of the mean) in the 95% equivalence test (i.e., 5% alpha). The estimated BF% and measured BF% across all protocols and the 90% CI for means of the estimated and measured BF% were obtained from a mixed ANOVA to control for participants' level clustering. It has also evaluated a novel approach to examine for 'significantly equivalence' rather than zero differences between different measurements [18].

3. Results

Table 2 summarizes particpants' basic characteristics including age, height, body weight and body mass index. The mean ± SD was calculated for age, height, weight, and BMI (20.01 ± 2.11 years, 174.81 ± 6.01 cm, 73.26 ± 13.60, and 23.91 ± 3.77 kg·m^{-2}).

Table 2. Characteristics and body composition of participants.

Variables	Males (n = 105)		
	Mean ± SD	Minimum	Maximum
Age (year)	20.01 ± 2.11	18.0	25.0
Height (cm)	174.81 ± 6.01	163.0	193.0
Weight (kg)	73.26 ± 13.60	58.0	150.0
BMI (kg·m^{-2})	23.91 ± 3.77	18.4	47.3

Note; BMI: Body Mass Index, SD: Standard Deviation.

Table 3 presents the correlation coefficients (r) between nine protocols and DEXA as a criterion method. Overall, the BF% measured by the nine protocols were highly correlated with the criterion measure (Max: r = 0.92, Min: r = 0.79). The strongest association was seen by protocol 8 (7-sites by Jackson & Pollock) and protocol 4 (3-sites by Pollock) (protocol 8: r = 0.92, protocol 4: r = 0.90), followed by protocols 7 (4-sites by Durnin & Wormersley, r = 0.87) and protocol 5 (4-sites NHCA Formula, r = 0.87).

Table 3. Correlations between DEXA and each protocol.

Variables	Mean ± SD	DEXA	P 1	P 2	P 3	P 4	P 5	P 6	P 7	P 8	P 9
DEXA	18.22 ± 8.04	1									
P 1	15.90 ± 6.64	0.843 **	1								
P 2	10.98 ± 4.41	0.851 **	0.755 **	1							
P 3	11.23 ± 4.74	0.866 **	0.772 **	0.863 **	1						
P 4	13.07 ± 4.73	0.901 **	0.843 **	0.865 **	0.870 **	1					
P 5	11.86 ± 4.39	0.873 **	0.788 **	0.826 **	0.953 **	0.898 **	1				
P 6	16.80 ± 8.65	0.770 **	0.693 **	0.837 **	0.946 **	0.814 **	0.903 **	1			
P 7	16.92 ± 4.04	0.877 **	0.826 **	0.834 **	0.814 **	0.916 **	0.822 **	0.781 **	1		
P 8	11.79 ± 4.52	0.926 **	0.825 **	0.885 **	0.954 **	0.937 **	0.965 **	0.900 **	0.927 **	1	
P 9	17.77 ± 3.80	0.790 **	0.737 **	0.725 **	0.803 **	0.844 **	0.772 **	0.749 **	0.842 **	0.837 **	1

Note; P: Protocol, ** Correlation is significant at the 0.01 level (two-tailed).

Figure 1 presents the MAPE for the nine protocols. Each protocol was calculated as the average absolute value of the errors based on DEXA, and the overall range of the error was from 20.0% to 38.8%. Protocols with the least errors were protocol 1 (1-point Bicep, 20.0%) and protocol 7 (4-sites by Durnin & Wormersley, 22.8%), followed by protocol 4 (3-sites by Pollock, 25.5%) and protocol 6 (4-sites by Forsyth-Sinning, 26.5%).

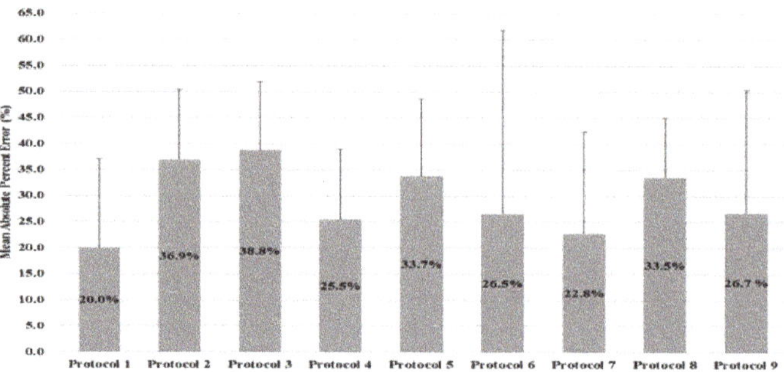

Figure 1. Mean absolute percentage error (MAPE) of body fat %.

A 1 × 9 repeated measures ANOVA with a post-hoc Bonferroni analysis showed that there was a significant difference (F [9, 1040] = 28.23; $p < 0.001$) between protocols, but no significant differences were observed between measured BF% and estimated BF% from Protocol 1 ($p = 0.133$), Protocol 6 ($p = 1.000$), Protocol 7 ($p = 1.000$), and Protocol 9 ($p = 1.000$). Figure 2 indicated the equivalence testing whether the BF% estimates from the nine protocols were equivalent to the estimate from the criterion measure (i.e., DEXA) and plots indicating the distribution of errors for all protocols. The calculated 90% CI for the estimates from the nine protocols were computed with the calculated equivalence zone for the DEXA. In particular, the estimated BF% from protocol 7 (4-sites by Durnin & Wormersley) and protocol 9 (9-sites Parllo) were significantly equivalent to the DEXA. This result showed that two different protocols (protocol 7 and protocol 9) were completely within the equivalence zone of the BF% measured by the DEXA (lower bound = 16.39%, upper bound = 20.03%).

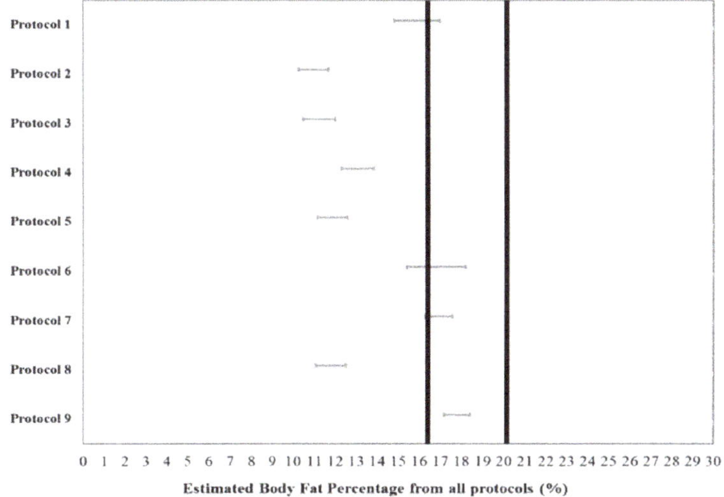

Figure 2. Dark lines mean equivalence zone (±10% of the mean), Grey lines are the 90% confidence interval for a mean of the estimated each protocol. Only protocol 7 and 9 are included within the equivalence zone.

Figure 3 illustrates the Bland-Altman plot for the extent of agreement between each protocol and the DEXA and assist with testing for proportional systematic bias in the estimates. The plots show mean bias, 95% limits of agreement, and line of best fit. The x-axis is the mean of two different measurements and the y-axis is the DEXA. The narrowest 95% limits of agreement of plots were protocol 8 (7-sites by Jackson & Pollock, difference = 16.4), followed by protocol 4 (3-sites by Pollock, difference = 16.7). Also, values of protocols were still higher for the protocol 1 (1-point Bicep, difference = 16.8), protocol 3 (3-sites by Jackson & Pollock, difference = 17.9), protocol 5 (4-sites NHCA Formula, difference = 18.4), protocol 2 (2-sites by A. W. Sloan, difference = 19.0), protocol 7 (4-sites by Durnin & Wormersley, difference = 19.0), protocol 9 (9-site Parllo, difference = 21.7), and protocol 6 (4-sites by Forsyth-Sinning, difference = 41.4). The clustering points of data in plots were indicated the mean for each protocol, compared to the DEXA and all protocols' slopes were observed as significant bias. The protocol 1 (intercept = 1.20, slope = −0.20, $p = 0.001$) and protocol 6 (intercept = 2.87, slope = −0.08, $p = 0.002$) tended to increase in a negative direction in comparison with the DEXA, however, another slopes for the fitted line showed a tendency to increase in a positive direction for protocol 2 (intercept = −1.90, slope = 0.62, $p = 0.001$), protocol 3 (intercept = −1.11, slope = 0.54, $p = 0.001$), protocol 4 (intercept = −3.35, slope = 0.54, $p = 0.000$), protocol 5 (intercept = −3.01, slope = 0.62, $p = 0.001$), protocol 7 (intercept = −11.0, slope = 0.70, p = 0.001), protocol 8 (intercept = −2.28, slope = 0.58, p = 0.000) and protocol 9 (intercept = −13.7, slope = 0.78, p = 0.001).

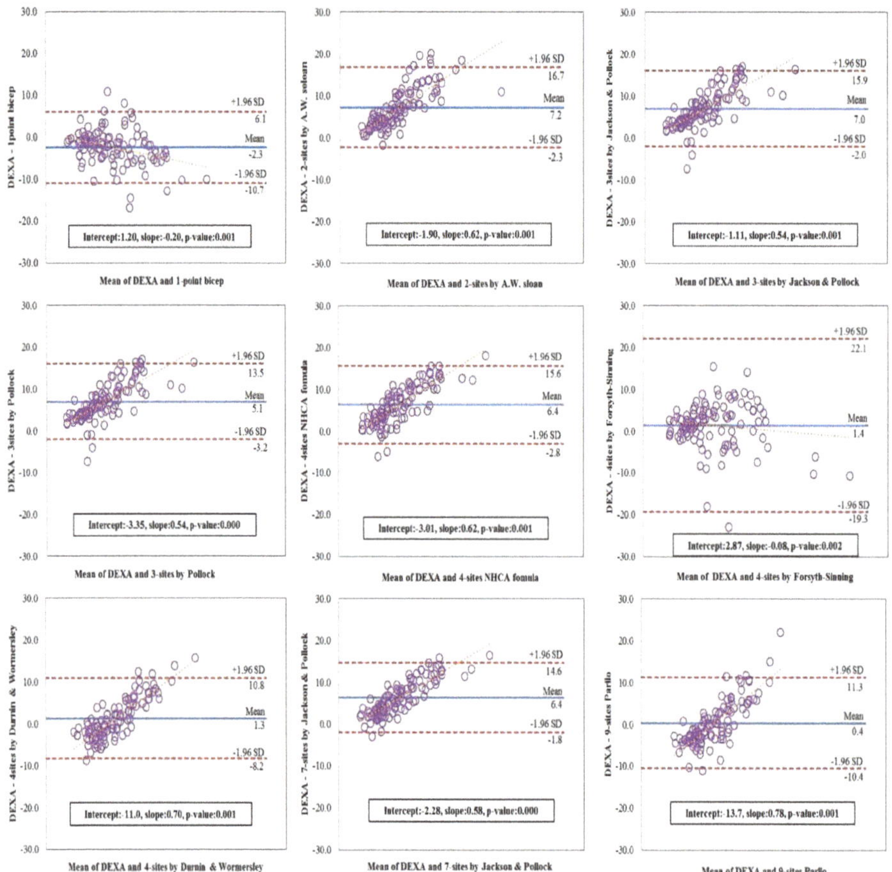

Figure 3. Bland-Altman plots for body fat percentage measured by each protocol.

4. Discussion

Skin fold thickness measurement is a cheap and non-invasive method to estimate body fat percentage and this method has been widely applied for all age goups. Different numbers of body site measurements with various equations has been introduced up to date, but investgators' technique plays an important role in obtaining accurate data. To minimize this potential error, advanced techniques using ultrasound such as a BodyMetrix™ BX-2000 has been introduced. This validation study compared the body fat percentage between nine different protocols built-in the BodyMetrix™ BX-2000 and DEXA measurements in collegiate male students. Overall, the estimated BF% from nine different protocols built-in the BodyMetrix™ BX-2000 were favorable compared with a gold standard method. Particulary, protocol 7 (4-sites by Durnin & Wormersley) and protocol 9 (9-site Parllo) have shown the most accurate measurements.

All nine protocols showed a relatively high correlation with the DEXA, in particular, protocol 8 (7-sites by Jackson & Pollock, r = 0.92) and protocol 4 (3-sites by Pollock, r = 0.90) showed the highest correlation. The consistent results have been identified in Kelly and colleagues' study where protocol 8 (7-sites by Jackson & Pollock) showed a high correlation with the DEXA (r = 0.92) in male collegiate students (n = 35, average BMI = 25.6 kg·m^{-2}) [19]. Also, there was a significat correlation between protocol 3 and BF% measured by DEXA in male collegiate student (n = 93, average BMI = 24 kg·m^{-2}) [20]. However, some differences between the results of MAPE and the correlations were found in the present study. Based on the results of MAPE values, the protocols with an error range below 30% were shown in protocol 1 (1-point Bicep, 20.0%), protocol 7 (4-sites by Durnin & Wormersley, 22.8%), protocol 4 (3-sites by Pollock, 25.5%), protocol 6 (4-sites by Forsyth-Sinning, 26.5%), and protocol 9 (9-site Parllo, 26.7%). There are some possible reasons that could explain the different results between correlation and MAPEs. For instance, the Pearson correlation was used to examine the two variables for each individual in a group, so it examines the group level association for each protocol in this study. When evaluating two variables' agreement using the correlation, two variables can be associated with each other. But the results may provide very different estimates at a group level. In addition, MAPEs delivers information regarding the individual agreement because it accounts for each participant's error while avoiding the cancellation of errors from under- and over-estimation. It means that the magnitude of error can be examined by calculating MAPE, but it does not quantify the overall direction (i.e., over- and under-estimation).

In contrast to the MAPE and correlation measures, Bland–Altman plots have clear utility to examine the visual inspection of under- and over-estimation between two variables, but a limitation is that they do not indicate a way to statistically evaluate the agreement. The information helps understand the nature and source of the proportional systematic biases, but it presents a significant challenge for drawing definitive conclusions based on the distributions in Bland-Altman plots. Standard inferential statistics such as t-tests and ANOVA are also often used to compare two measures and assess group agreement. However, these statistical tests are designed to test for differences rather than agreement.

Therefore, the present study used a more appropriate analytical method such as equivalence testing to identify validity. Some studies demonstrated that equivalence testing is a more powerful and reliable method for comparing different assessment measures and evaluating agreement among measurements [21–23]. The remarkable findings from this study were that the results of equivalence testing were some seemingly discrepant results based on the results of correlation, MAPEs, and inferential statistics (i.e., repeated-measures ANOVA. Protocol 8 and Protocol 4, for example, showed a high correlation between the BF% estimated by the ultrasound device compared to BF% measured by the DEXA, and Protocol 1, 6, 7, and 9 did not show the statistical differences compared to measured BF% by the DEXA. However, the results of protocol 7 (4-sites by Durnin & Wormersley) and protocol 9 (9-site Parllo) located completely within the proposed equivalence zone (e.g., ±10% of the mean) in the 95% equivalence test, and it means that protocol 7 and protocol 9 had confirmed the validity of the portable ultrasound device.

There are some strength and limitations in the present study. First of all, this is the first study that investigate the validity of the total nine protocols built-in the BodyMetrix™ BX-2000 for estimating BF% through advanced statistical testing, "equivalence testing". Moreover, this study also exmined that equivalence testing is a reliable statistical method rather than other methods (i.e., correlation, MAPE) when comparing the validity of different measuring devices. Therefore, the study demonstrated that protocol 7 (4-sites by Durnin & Wormersley) and protocol 9 (9-site Parllo) were more adaptable for measuring BF% of male college students with a normal BMI. Also, considering the measurement efficiency, using the values measured Protocol 7 (4-sites by Durnin & Wormersley) would be more compatible to evaluate BF% because it requires fewer measurements. The study provided new insights about the BodyMetrix™ BX-2000, but it does have some limitations. The age range of sample participants is limited to college students, and gender is also limited to only males. Therefore, it is difficult to generalize findings in this study to other age groups or females. Also, our study did not assess the reliability of each protocol even if poor reliability can negatively impact the validity, and except the protocol 1 and protocol 6, proportional systemic biases were observed in Bland-Altman. In terms of equivalent testing in this study, ±10% of the mean of the DEXA was used as a lower and upper boundary of the equivalence zone. However, more supportive research is clearly needed to compare results with diverse populations and criterion measures.

5. Conclusions

The present study evaluated the validity of a commercially available ultrasound device to measure BF%. The study shows that a small, portable, and easy-to-use ultrasound system allows reasonably good estimates of BF%. Furthermore, measuring BF% with the ultrasound BodyMetrix™ BX-2000 rather than measuring it by the expensive and laboratory-based DEXA is shown comparable estimation of BF%. Reasonable estimates of BF% enable health and fitness practitioners to provide better care to their clients or patients regarding healthy body composition.

Author Contributions: Data curation S.K. and J.-M.L.; Formal analysis, S.K., J.-H.P., Y.I.K. and J.-M.L.; Investigation, S.K., Y.I.K. and J.-M.L.; Methodology, S.K., M.-W.S., H.C.J., Y.I.K. and J.-M.L.; Project administration, S.K.; Writing—original draft, S.K.; Writing—review & editing, J.-H.P., M.-W.S., H.C.J. and J.-M.L. All authors have read and agreed to the published version of the manuscript.

Funding: This work was supported by the 2016 Research Fund of the Yongin University.

Acknowledgments: The authors would like to thank all participants who participated in this study.

Conflicts of Interest: The authors no conflict of interest to declare.

References

1. World Health Organization. *Obesity and Overweight*; World Health Organization: Geneva, Switzerland, 2017.
2. Lee, A.; Cardel, M.; Donahoo, W.T. Social and Environmental Factors Influencing Obesity. In *Endotext [Internet]*; MDText. com, Inc.: South Dartmouth, MA, USA, 2019.
3. Valle, M.; Gascón, F.; Martos, R.; Ruz, F.J.; Bermudo, F.; Morales, R.; Cañete, R. Metabolic cardiovascular syndrome in obese prepubertal children: The role of high fasting insulin levels. *Metab. Clin. Exp.* **2002**, *51*, 423–428. [CrossRef] [PubMed]
4. Kumari, M.; Kozyrskyj, A.L. Gut microbial metabolism defines host metabolism: An emerging perspective in obesity and allergic inflammation. *Obes. Rev. Off. J. Int. Assoc. Study Obes.* **2017**, *18*, 18–31. [CrossRef] [PubMed]
5. Thiebaud, R.S.; Abe, T.; Loenneke, J.P.; Fujita, E.; Akamine, T. Body fat percentage assessment by ultrasound subcutaneous fat thickness measurements in middle-aged and older adults. *Clin. Nutr.* **2019**, *38*, 2659–2667. [CrossRef] [PubMed]
6. Hoffman, D.J.; Policastro, P.; Quick, V.; Lee, S.-K. Changes in body weight and fat mass of men and women in the first year of college: A study of the "freshman 15". *J. Am. Coll. Health* **2006**, *55*, 41–46. [CrossRef] [PubMed]
7. Gropper, S.S.; Simmons, K.P.; Connell, L.J.; Ulrich, P.V. Changes in body weight, composition, and shape: A 4-year study of college students. *Appl. Physiol. Nutr. Metab. Physiol. Appl. Nutr. Metab.* **2012**, *37*, 1118–1123. [CrossRef] [PubMed]
8. Glaner, M.F.; Lima, W.A.; Borysiuk, Z. Body fat deposition and risk factors of cardiovascular diseases in men. *Hum. Mov.* **2010**, *11*, 45–50.
9. Gómez-Ambrosi, J.; Silva, C.; Galofré, J.C.; Escalada, J.; Santos, S.; Gil, M.J.; Valentí, V.; Rotellar, F.; Ramírez, B.; Salvador, J. Body adiposity and type 2 diabetes: Increased risk with a high body fat percentage even having a normal BMI. *Obesity* **2011**, *19*, 1439–1444. [CrossRef] [PubMed]
10. Loenneke, J.P.; Hirt, K.M.; Wilson, J.M.; Barnes, J.T.; Pujol, T.J. Predicting body composition in college students using the womersley and durnin body mass index equation. *Asian J. Sports Med.* **2013**, *4*, 153. [CrossRef] [PubMed]
11. Morrow, J.R., Jr.; Mood, D.; Disch, J.; Kang, M. *Measurement and Evaluation in Human Performance, 5E*; Human Kinetics: Champaign, IL, USA, 2015.
12. Smith-Ryan, A.E.; Blue, M.N.; Trexler, E.T.; Hirsch, K.R. Utility of ultrasound for body fat assessment: Validity and reliability compared to a multicompartment criterion. *Clin. Physiol. Funct. Imaging* **2018**, *38*, 220–226. [CrossRef] [PubMed]

13. Lyon, J.; Drew, R.; MacRae, H. Comparison of skinfold thickness measures with ultrasound imaging to determine body composition. In Proceedings of the 26th Annual Meeting of the Southwest Chapter of the American College of Sports Medicine, San Diego, CA, USA, 10–11 November 2006.

14. Ulbricht, L.; Neves, E.B.; Ripka, W.L.; Romaneli, E.F. Comparison between body fat measurements obtained by portable ultrasound and caliper in young adults. In Proceedings of the 2012 Annual International Conference of the IEEE Engineering in Medicine and Biology Society, San Diego, CA, USA, 28 August–1 September 2012; IEEE: San Diego, CA, USA, 2012; pp. 1952–1955.

15. Kang, S.; Lee, M.; Kim, Y. Feasibility of Ultrasound Measuring Equipment by Body Fat Percentage Examination. *J. Asian Soc. Health Exerc.* **2019**, *1*, 19–29.

16. Johnson, K.E.; Naccarato, I.A.; Corder, M.A.; Repovich, W.E. Validation of three body composition techniques with a comparison of ultrasound abdominal fat depths against an octopolar bioelectrical impedance device. *Int. J. Exerc. Sci.* **2012**, *5*, 205. [PubMed]

17. Loenneke, J.P.; Barnes, J.T.; Wagganer, J.D.; Wilson, J.M.; Lowery, R.P.; Green, C.E.; Pujol, T.J. Validity and reliability of an ultrasound system for estimating adipose tissue. *Clin. Physiol. Funct. Imaging* **2014**, *34*, 159–162. [CrossRef] [PubMed]

18. Hauck, W.W.; Anderson, S. A new statistical procedure for testing equivalence in two-group comparative bioavailability trials. *J. Pharmacokinet. Biopharm.* **1984**, *12*, 83–91. [CrossRef] [PubMed]

19. Johnson, K.E.; Miller, B.; Juvancic-Heltzel, J.A.; Agnor, S.E.; Kiger, D.L.; Kappler, R.M.; Otterstetter, R. Agreement between ultrasound and dual-energy X-ray absorptiometry in assessing percentage body fat in college-aged adults. *Clin. Physiol. Funct. Imaging* **2014**, *34*, 493–496. [CrossRef] [PubMed]

20. Pineau, J.-C.; Filliard, J.R.; Bocquet, M. Ultrasound techniques applied to body fat measurement in male and female athletes. *J. Athl. Train.* **2009**, *44*, 142–147. [CrossRef] [PubMed]

21. Staudenmayer, J.; Zhu, W.; Catellier, D.J. Statistical considerations in the analysis of accelerometry-based activity monitor data. *Med. Sci. Sports Exerc.* **2012**, *44*, S61–S67. [CrossRef] [PubMed]

22. Hopkins, W.; Marshall, S.; Batterham, A.; Hanin, J. Progressive statistics for studies in sports medicine and exercise science. *Med. Sci. Sports Exerc.* **2009**, *41*, 3. [CrossRef] [PubMed]

23. Dixon, P.M.; Saint-Maurice, P.F.; Kim, Y.; Hibbing, P.; Bai, Y.; Welk, G.J. A primer on the use of equivalence testing for evaluating measurement agreement. *Med. Sci. Sports Exerc.* **2018**, *50*, 837. [CrossRef] [PubMed]

Publisher's Note: MDPI stays neutral with regard to jurisdictional claims in published maps and institutional affiliations.

Article

Analysis of Self-Concept in Adolescents before and during COVID-19 Lockdown: Differences by Gender and Sports Activity

Gabriel González-Valero [1], Félix Zurita-Ortega [1], David Lindell-Postigo [2], Javier Conde-Pipó [1], Wilhelm Robert Grosz [3] and Georgian Badicu [3,*]

1   Department of Didactics of Musical, Plastic and Corporal Expression, University of Granada, Campus de Cartuja, s/n 18071 Granada, Spain; ggvalero@ugr.es (G.G.-V.); felixzo@ugr.es (F.Z.-O.); javiconde@correo.ugr.es (J.C.-P.)
2   Novaschool Sunland International, Estación de Cártama-Málaga, Carretera de Cártama Estación a Pizarra, s/n 29580 Cártama, Spain; dlindell@novaschool.es
3   Department of Physical Education and Special Motricity, Faculty of Physical Education and Mountain Sports, Transilvania University of Brasov, 500068 Brasov, Romania; wilhelm.grosz@unitbv.ro
*   Correspondence: georgian.badicu@unitbv.ro; Tel.: +40-769-219-271

Received: 3 September 2020; Accepted: 19 September 2020; Published: 21 September 2020

Abstract: An appeal has been issued to the scientific community to investigate physical, mental and emotional states, and pro-social behaviours during the COVID-19 pandemic. Hence, this study aims to investigate adolescents' self-concept before and during a lockdown period in relation to gender and type/amount of physical activity or sports. The pre-lockdown sample of 366 adolescents were aged 13–17 years (M = 15.51 ± 0.65), of whom 192 (52.5%) were females and 174 (47.5%) were males. During the lockdown, the age range of the sample was 13–17 years (M = 14.57 ± 1.47), of whom 82 (60.3%) were females, and 54 (39.7%) were males. The Form-5 Self-concept Questionnaire (AF-5) was used to measure adolescents' self-concept. There was a reduction in adolescents' overall self-concept during the COVID-19 pandemic, which was positively associated with emotional well-being, with family and peers being essential factors in the development of an appropriate self-concept. Furthermore, girls' self-concept, especially academic self-concept, was higher than that of boys during the lockdown. However, both physical and emotional self-concept were higher for boys than girls before the COVID-19 lockdown, although no differences were found during the lockdown. The findings reveal that physical activity was positively correlated to self-concept before and during the COVID-19 lockdown.

Keywords: self-concept; physical activities; lockdown; COVID-19; adolescents

---

1. Introduction

Every country negatively affected by a disaster or a pandemic considers adolescents, the elderly and disabled people to be the main at-risk population groups [1–3]. The coronavirus SARS-CoV-2 infection, which leads to the disease called COVID-19 [4], caused an international state of health emergency and global pandemic [5]. This disease features respiratory infections, which directly affect the elderly [4,6]. However, adolescents are indirectly affected due to social distancing, educational and recreational measures adopted by the government [7,8]. The pandemic has had economic impacts and caused social disruptions [9,10].

Furthermore, COVID-19 globally jeopardises people's mental health since it increases stress, anxiety, depression and negative social behaviour [9–11]. Therefore, adolescents are in danger of experiencing negative consequences for both mental and physical health as a result of pandemics and disasters [2,12]. Likewise, COVID-19 is affecting aspects of daily life, such as educational, social and leisure activities, which present both familial and emotional challenges [7,8,13,14]. Moreover, psychological needs, such as self-fulfilment, self-esteem and affective relationships, take on greater importance once both physiological and security requirements are met [15].

It should be noted that self-concept, which is understood as one's perception about oneself or the general opinion about self-esteem [16], may buffer people's psychological distress [17,18]. Hence, self-concept is especially important in adolescence, since everyday emotions and feelings are essential in personal development, which is subjective and changes according to external factors and new contexts of life [19]. As a matter of fact, self-concept represents a protective factor against disruptive behaviour, enhancing both mental health and positive peer relationships [20]. Consequently, the psychological construction of a positive self-concept in students during the school years produces successful socio-emotional situations and educational settings [21].

The literature has shown gender differences in well-being and self-concept. Females' well-being and self-concept are related to life satisfaction and happiness, whereas males' well-being and self-concept are related to feelings of achievement [22,23]. The effect of the COVID-19 pandemic has not been studied and no conclusive results have therefore been drawn. The need to study the association between self-concept and physical activity during this period lies in the importance of children's development, with the physical practice being a means of improving mental processes and socialisation in children [19]. Adolescents who habitually participate in physical and sports activities have a better self-concept, which is the social motor that drives better academic performance and helps their relations with peers [24].

Hence, this construct should be considered in youths and their relatives who usually experience post-traumatic symptoms under pandemic or emergency situations [9,11,25], which may trigger negative mental health consequences, disruptive behaviour and a low self-concept [26–29]. As a consequence of the coronavirus pandemic, an appeal has been issued to the scientific community to pay attention to and analyse the physical, mental and emotional conditions as well as people's pro-social behaviour [13,30,31]. In relation to the study problem, the following research questions were suggested: (a) are there differences in adolescents' levels of self-concept before and during COVID-19? (b) are there differences between boys and girls? (c) does being physically active or inactive influence adolescents' self-concept before and during COVID-19? Therefore, this study aims to examine the level of adolescents' self-concept before lockdown and during lockdown as regards gender and physical activity, for boys and girls, and for those who are physically active and those who are not.

2. Materials and Methods

2.1. Design and Participants

This study compared the self-concept of two groups of adolescents, one pre-lockdown and one during lockdown. Convenience sampling was used to select participants, in which adolescents were asked to participate before and during the lockdown, so different samples were evaluated at two different times. As regards this selection criterion, 72.9% (n = 366) of participants were assessed before the COVID-19 lockdown and 27.1% (n = 136) of participants were assessed during that period. Both groups were equivalent in all respects, except for the lockdown situation. The age range of the sample before the lockdown was 13–17 years (M = 15.51 ± 0.65), while females accounted for 52.5% (n = 192), males accounted for 47.5% (n = 174). During the lockdown, the age range of the sample was 13–17 years (M = 14.57 ± 1.47); females constituted 60.3% (n = 82) of the total, whereas males accounted for 39.7% (n = 54). In order to obtain a representative sample (error at 0.05; C.I. = 95%), stratification and proportionality techniques were used when establishing the groups.

2.2. Instruments and Variables

Ad-hoc questionnaire. This instrument was created by the researchers and was used to collect socio-demographic and physical and sports data. Thus, the data collected were participants' gender and age, moment (in lockdown or not), and type of physical or sport activities practised according to the adapted classification of Castro-Sánchez et al. [32]. This classification involves the following categorisation: 'None', 'Non-contact individual sports' (NCIS), 'Contact individual sports' (CIS), 'Non-contact team sports' (NCTS) and 'Contact team sports' (CTS). Furthermore, based on this classification, adolescents were classified as "physically active" and "physically inactive".

Form-5 Self-concept Questionnaire (AF-5). This questionnaire assesses one's perception about one's self-concept, which is based on the theory stated by Shavelson, Hubner and Stanton [33], and was created and validated to Spanish by García and Musitu [34]. It consists of 30 items, which use five-points Likert responses ranging from 'Never' to 'Always'. Item summation allows us to establish a general measurement of this construct, as well as to group self-concept into five dimensions: academic self-concept (A-SC; items 1, 6, 11, 16, 21 and 26), social self-concept (S-SC; items 2, 7, 12, 17, 22 and 27), emotional self-concept (E-SC; items 3, 8, 13, 18, 23 and 28), family self-concept (F-SC; items 4, 9, 14, 19, 24 and 29) and physical self-concept (P-SC; items 5, 10, 15, 20, 25 and 30). Cronbach's alpha in this study ($\alpha = 0.809$) was similar to García and Musitu [34] research ($\alpha = 0.810$). Reliabilities of each dimension of self-concept were as follows: A-SC ($\alpha = 0.853$), S-SC ($\alpha = 0.784$), E-SC ($\alpha = 0.756$), F-SC ($\alpha = 0.706$) and P-SC ($\alpha = 0.801$).

2.3. Procedure

Firstly, researchers explored the range of ways to contact the population. Adolescents evaluated before the lockdown were asked thenceforth to participate in this investigation via an informative letter, which was created by the Body language department of the University of Granada and delivered through their schools. A meeting with principals was arranged afterwards, in which researchers handed over to principals some hard copies of evaluation instruments and the informative letter that needed to be delivered to adolescents' families. That letter detailed the objectives and the nature of the research, explained the voluntariness of the participation and requested informed consent. Data collection was conducted in school time during Physical Education lessons in the presence of researchers and teachers, in order to solve any doubt and to ensure a correct completion, not occurring any incidence during the process. The procedure was similar with participants evaluated during the lockdown, although protocols were done online, questionnaires were distributed via social media, and contact with families was done through schools' communication channels (e.g., email, blogs or *Telegram*). Anonymity was ensured in both processes, and researchers also certified that data would be used for scientific purposes. The study was conducted in full compliance with the principles expressed in the Declaration of Helsinki and was approved by the Scientific Ethical Committee of the research team's university (1230/CEIH/2020). Lastly, researchers had to invalidate 37 questionnaires due to incorrect completion.

2.4. Data Analysis

Descriptive analysis for variables in this study was performed, calculating mean values (M), standard deviation (SD) and frequencies (%). Normality and homogeneity of variance for every variable were analysed through the Kolmogorov–Smirnov test. Researchers performed the independent Student's t-test to estimate differences among variables and performed the Bonferroni post-hoc test to determine the one-way variance (ANOVA) with one group, determined by Pearson's chi-squared test. Pearson bivariate correlation was calculated to establish an association among mean values, calculating a significance level of $p \leq 0.05$ and $p \leq 0.01$. Lastly, the magnitude of difference in effect size (ES) was obtained with Cohen's *d* index [35], which is interpreted as null (0–0.19), small (0.20–0.49), medium (0.50–0.7), and large ($\geq$0.80) [36]. Data were analysed using IBM SPSS® version 25.0 (IBM Corp,

Armonk, NY, USA). GraphPad Prism 8 (GraphPad Prism Software Inc., San Diego, CA, USA) was used to produce figures. Lastly, 37 questionnaires were invalidated for incorrect completion, of which 26 were from individuals before the lockdown and 11 were during the COVID-19 lockdown.

3. Results

Table 1 presents values of self-concept regarding lockdown during the COVID-19 pandemic ($p \leq 0.05$). The highest values of total self-concept were before the lockdown (M = 3.47 ± 0.52; d = 0.253), presenting a positive association with emotional self-concept (r = 0.375**). This association was not present during the lockdown (r = 0.090). The social self-concept (M = 3.70 ± 0.75; d = 0.401) was influenced by its correlation with emotional self-concept (r = 0.200**). The emotional self-concept (M = 3.02 ± 0.78; d = 0.482) had a positive association with family self-concept (r = 0.143**) in both moments. Likewise, there existed a negative association between physical and emotional dimensions of self-concept during the lockdown (r = −0.424**). Furthermore, while individuals evaluated before the lockdown presented higher values in family self-concept (M = 4.02 ± 0.83; d = 0.643), during the lockdown presented them in the academic dimension (M = 3.85 ± 0.70; d = 0.705).

Table 1. Self-concept before and duringCOVID-19 pandemic.

Variables	Lockdown	M	SD	(2)	(3)	(4)	(5)	(6)	Levene F	Levene Sig.	T-Test T	T-Test Sig.	ES (d)	IC 95%
G-SC (1)	Yes	3.35	0.32	0.744 **	0.695 **	0.090	0.591 **	0.637 **	24.222	0.000	−3.003	0.003	0.253	[0.056; 0.450]
	No	3.47	0.52	0.612 **	0.754 **	0.375 **	0.749 **	0.717 **						
A-SC (2)	Yes	3.85	0.70	1	0.499 **	−0.210 *	0.266 **	0.413 **	9.695	0.002	7.836	0.000	0.705	[0.504; 0.907]
	No	3.25	0.90	1	0.247 **	−0.125 *	0.371 **	0.378 **						
S-SC (3)	Yes	3.43	0.40		1	−0.130	0.337 **	0.398 **	58.009	0.000	−5.149	0.000	0.401	[0.202; 0.599]
	No	3.70	0.75		1	0.200 **	0.499 **	0.523 **						
E-SC (4)	Yes	2.66	0.65			1	0.027	−0.424 **	7.036	0.008	−5.239	0.000	0.482	[0.283; 0.681]
	No	3.02	0.78			1	0.143 **	0.046						
F-SC (5)	Yes	3.54	0.45				1	0.214 *	71.469	0.000	−8.172	0.000	0.643	[0.442; 0.844]
	No	4.02	0.83				1	0.369 **						
P-SC (6)	Yes	3.29	0.76					1	0.196	0.658	−0.909	0.364	0.090	[−0.107; 0.287]
	No	3.36	0.78					1						

Note 1. General self-concept (G-SC); academic self-concept (A-SC); social self-concept (S-SC); emotional self-concept (E-SC); family self-concept (F-SC); physical self-concept (P-SC). Note 2. Significative correlation at 0.05 (*); significative correlation at 0.01 (**).

Figure 1 shows the relation between self-conceptandthe lockdown. The general self-concept before the lockdown was higher than during this period. The same happened with the S-SC, E-SC and F-SC. In contrast, the A-SC was higher during the lockdown period. No statistically significant results were found for the P-SC.

Table 2 presents differences between gender and self-concept before the COVID-19 lockdown. Girls had a greater academic self-concept (M = 3.39 ± 0.90; d = 0.339) with a highly significant difference (p = 0.001). However, among the boys, a negative association was shown between that dimension and emotional self-concept (r = −0.287**). There was a significant difference ($p \leq 0.05$) in some components before the lockdown. Boys had specifically higher values in the P-SC (M = 3.44 ± 0.77; d = 0.320) and E-SC (M = 3.14 ± 0.73; d = 0.298). However, girls' E-SC was positively correlated with the F-SC (r = 0.154*).

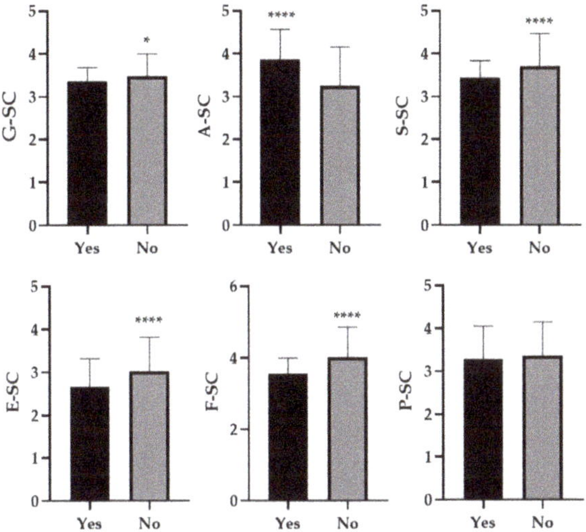

Figure 1. Self-concept and dimensions according to the period of lockdown. Note 1. Self-concept (SC); general self-concept (G-SC); academic self-concept (A-SC); social self-concept (S-SC); emotional self-concept (E-SC); family self-concept (F-SC); physical self-concept (P-SC). Note 2. Lockdown (Yes); nolockdown (No). Note 3. $p \leq 0.05$ (*); $p \leq 0.0001$ (****).

Table 2. Gender-based differences in self-concept before COVID-19 lockdown.

Lockdown	Variables	Sex	M	SD	(2)	(3)	(4)	(5)	(6)	Levene		T-Test		ES (d)	IC 95%
										F	Sig.	T	Sig.		
SC before lockdown	G-SC (1)	B	3.48	0.48	0.526 **	0.768 **	0.288 **	0.730 **	0.734 **	2.598	0.108	0.379	0.705	0.039	[−0.167; 0.244]
		G	3.46	0.55	0.702 **	0.747 **	0.439 **	0.763 **	0.709 **						
	A-SC (2)	B	3.09	0.87	1	0.165 *	−0.287 **	0.220 **	0.386 **	0.026	0.873	−3.240	0.001	0.339	[0.132; 0.545]
		G	3.39	0.90	1	0.342 **	0.044	0.493 **	0.413 **						
	S-SC (3)	B	3.74	0.76		1	0.196 **	0.497 **	0.523 **	0.032	0.858	0.851	0.395	0.093	[−0.113; 0.298]
		G	3.67	0.75		1	0.197 **	0.505 **	0.519 **						
	E-SC (4)	B	3.14	0.73			1	0.140	−0.065	2.199	0.139	2.830	0.005	0.298	[0.091; 0.504]
		G	2.91	0.81			1	0.154 *	0.112						
	F-SC (5)	B	4.00	0.79				1	0.376 **	3.381	0.067	−0.451	0.652	0.048	[−0.157; 0.253]
		G	4.04	0.88				1	0.371 **						
	P-SC (6)	B	3.44	0.77					1	0.024	0.877	1.789	0.004	0.320	[0.114; 0.527]
		G	3.19	0.79					1						

Note 1. Boy (B); girl(G). Note 2. Self-concept (SC); general self-concept (G-SC); academic self-concept (A-SC); social self-concept (S-SC); emotional self-concept (E-SC); family self-concept (F-SC); physical self-concept (P-SC). Note 3. Significative correlation at 0.05 (*); significative correlation at 0.01 (**).

Figure 2 shows the relationship between self-concept and gender-based differences before lockdown. Girls had greater levels ofA-SC, although boys had higher E-SC and P-SC. No statistically significant results ($p \geq 0.05$) were detected for the G-SC, S-SC and F-SC.

Table 3 presents differences according to gender regarding self-concept during the COVID-19 lockdown. There were also significant differences ($p \leq 0.05$) during lockdown. Females had greater values in A-SC (M = 3.97 ± 0.70; d = 0.421), while for males, A-SC had a negative association with E-SC (r = −0.281*) and a positive correlation with F-SC (r = 0.414**). Significant differences were shown in G-SC (p = 0.027), where girls had greater values than boys (M = 3.38 ± 0.31; d = 0.346).

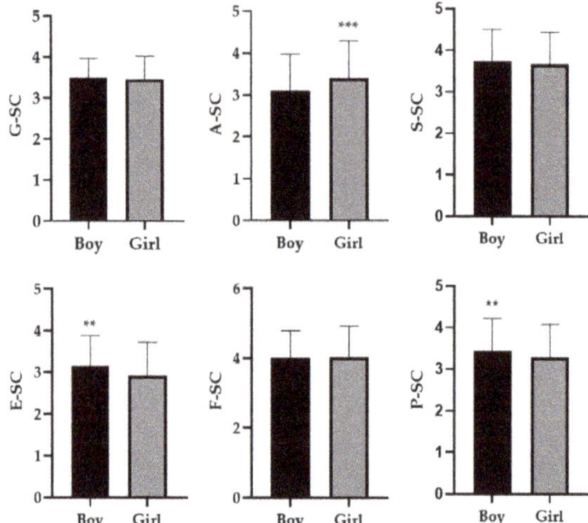

Figure 2. Gender-based differences in self-concept before COVID-19 lockdown. Note 1. Self-concept (SC); general self-concept (G-SC); academic self-concept (A-SC); social self-concept (S-SC); emotional self-concept (E-SC); family self-concept (F-SC); physical self-concept (P-SC). Note 2. $p \leq 0.01$ (**); $p \leq 0.001$ (***).

Table 3. Gender-based differences in self-concept during COVID-19 lockdown.

Lockdown	Variables	Sex	M	SD	(2)	(3)	(4)	(5)	(6)	Levene F	Levene Sig.	T-Test T	T-Test Sig.	ES (*d*)	IC 95%
SC during lockdown	G-SC (1)	M	3.27	0.33	0.837 **	0.706 **	−0.024	0.614 **	0.710 **	0.520	0.729	−2.092	0.027	0.346	[0.001; 0.692]
		F	3.38	0.31	0.682 **	0.695 **	0.170	0.576 **	0.582 **						
	A-SC (2)	M	3.68	0.67	1	0.596 **	−0.281 *	0.414 **	0.602 **	0.335	0.564	−2.375	0.019	0.421	[0.074; 0.768]
		F	3.97	0.70	1	0.456 **	−0.170	0.175	0.289 **						
	S-SC (3)	M	3.43	0.39		1	−0.231	0.320 *	0.484 **	0.014	0.905	−0.040	0.968	0.001	[−0.343; 0.343]
		F	3.43	0.41		1	−0.072	0.349 **	0.339 **						
	E-SC (4)	M	2.67	0.63			1	0.009	−0.484 **	0.472	0.493	0.185	0.853	0.031	[−0.313; 0.374]
		F	2.65	0.67			1	0.038	−0.387 **						
	F-SC (5)	M	3.54	0.46				1	0.144	0.147	0.702	−0.109	0.913	0.001	[−0.343; 0.343]
		F	3.54	0.44				1	0.223 *						
	P-SC (6)	M	3.27	0.81					1	0.013	0.908	−0.225	0.822	0.040	[−0.304; 0.383]
		F	3.30	0.72					1						

Note 1. Male/Boy (M); Female/Girl (F). Note 2. Self-concept (SC); general self-concept (G-SC); academic self-concept (A-SC); social self-concept (S-SC); emotional self-concept (E-SC); family self-concept (F-SC); physical self-concept (P-SC). Note 3. Significative correlation at 0.05 (*); significative correlation at 0.01 (**).

Figure 3 shows the relationships between self-concept and gender-based differences during lockdown. Especially the girls had a greater G-SC and A-SC than boys. No statistically significant results ($p \geq 0.05$) were found for the other dimensions.

Table 4 presents the association between practising physical activities and self-concept before the lockdown ($p \leq 0.05$). Individuals who were not involved in any physical activity had the lowest values of G-SC (M = 3.30 ± 0.49) compared to NCIS (M = 3.58 ± 0.52; d = 0.554) and STS participants (M = 3.59 ± 0.53; d = 0.574). Individuals who practiced NCIS had the highest values in A-SC (M = 3.47 ± 0.87; d = 0.905) compared to individuals who did not practice any physical activity (M = 2.73 ± 0.76). The practice of NCIS or CTS was associated to higher values of S-SC (M = 3.83 ± 0.68; d = 0.451 and M = 3.83 ± 0.80; d = 0.419), F-SC (M = 4.11 ± 0.81; d = 0.364 and M = 4.26 ± 0.74; d = 0.549) and P-SC (M = 3.47 ± 0.72; *d* = 0.522 and M = 3.57 ± 0.77; *d* = 0.596), compared to not practising any physical activity or sport.

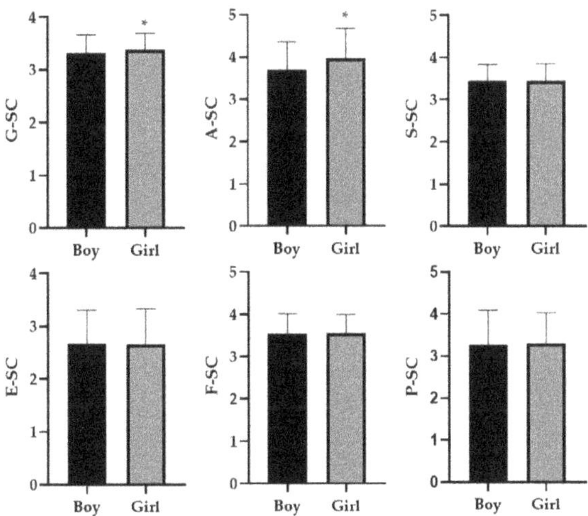

Figure 3. Gender-based differences in self-concept during COVID-19 lockdown. Note 1.Self-concept (SC); general self-concept (G-SC); academic self-concept (A-SC); social self-concept (S-SC); emotional self-concept (E-SC); family self-concept (F-SC); physical self-concept (P-SC). Note 2. $p \leq 0.05$ (*).

Table 4. Practice of physical or sport activities and self-concept before the COVID-19 lockdown.

Variable	Sport	Mean	SD	F	Sig.	ES (*d*)	IC 95%
G-SC	NS	3.30	0.49	5.826	$p \leq 0.05$ [a,b]	0.554 [a] 0.574 [b]	[0.300; 0.808] [a] [0.278; 0.871] [b]
	CIS	3.35	0.35				
	NCIS	3.58	0.52				
	CTS	3.59	0.53				
	NCTS	3.46	0.52				
A-SC	NS	2.73	0.76	7.492	$p \leq 0.05$ [a]	0.905 [a]	[0.644; 1.167] [a]
	CIS	3.16	0.88				
	NCIS	3.47	0.87				
	CTS	3.14	0.98				
	NCTS	3.21	0.74				
S-SC	NS	3.50	0.78	3.684	$p \leq 0.05$ [a,b]	0.451 [a] 0.419 [b]	[0.149; 0.737] [a] [0.125; 0.713] [b]
	CIS	3.58	0.62				
	NCIS	3.83	0.68				
	CTS	3.83	0.80				
	NCTS	3.69	0.72				
E-SC	NS	2.96	0.77	1.552	$p \geq 0.05$	NP	NP
	CIS	3.31	0.81				
	NCIS	3.01	0.82				
	CTS	3.17	0.77				
	NCTS	2.88	0.63				

Table 4. *Cont.*

Variable	Sport	Mean	SD	F	Sig.	ES (*d*)	IC 95%
F-SC	NS	3.80	0.89	4.210	$p \leq 0.05$ [a,b]	0.364 [a] 0.549 [b]	[0.113; 0.615] [a] [0.253; 0.845] [b]
	CIS	3.83	0.72				
	NCIS	4.11	0.81				
	CTS	4.26	0.74				
	NCTS	4.06	0.81				
P-SC	NS	3.10	0.80	5.562	$p \leq 0.05$ [a,b]	0.522 [a] 0.596 [b]	[0.299; 0.806] [a] [0.299; 0.892] [b]
	CIS	3.30	0.51				
	NCIS	3.47	0.72				
	CTS	3.57	0.77				
	NCTS	3.46	0.86				

Note 1. General self-concept (G-SC); academic self-concept (A-SC); social self-concept (S-SC); emotional self-concept (E-SC); family self-concept (F-SC); physical self-concept (P-SC). Note 2. No sport(NS); non-contact individual sport (NCIS); contact individual sport (CIS); non-contact team sport (NCTS); contact team sport (CTS). Note 3. Differences between NP and NCIS ([a]); differences between NP and CTS ([b]).

Figure 4 compares groups of physically active and inactive individuals before the lockdown. Being physically active before lockdown was associated with higher levels of G-SC, A-SC, S-SC, F-SC and P-SC. However, no statistically significant differences ($p \geq 0.05$) were found for the E-SC dimension.

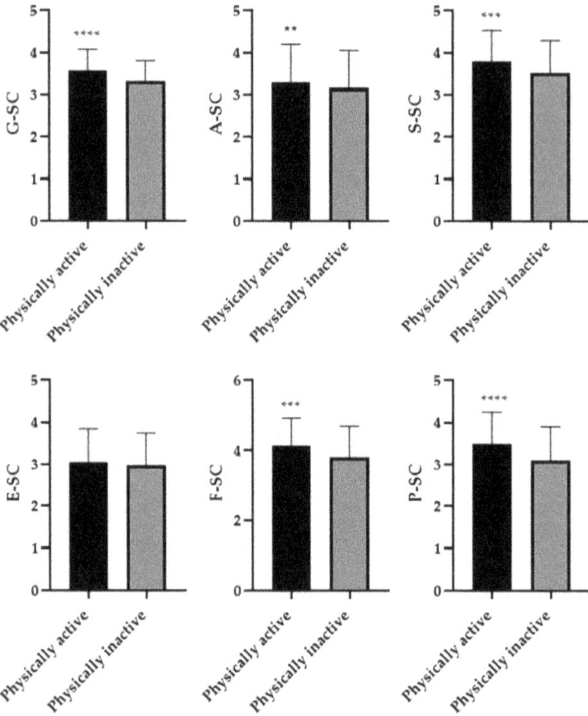

Figure 4. Self-concept and dimensions concerning physically active and inactive adolescents before the lockdown. Note 1. Self-concept (SC); general self-concept (G-SC); academic self-concept (A-SC); social self-concept (S-SC); emotional self-concept (E-SC); family self-concept (F-SC); physical self-concept (P-SC). Note 2. $p \leq 0.01$ (**); $p \leq 0.001$ (***); $p \leq 0.0001$ (****).

For the association between the practice of physical or sport activities and self-concept during the lockdown (Table 5), significant differences were found ($p \leq 0.05$). The subjects who performed NCIS activities (M = 3.48 ± 0.26; d = 0.827) were those who had higher values in G-SC compared to non-realised physical activity (M = 3.22 ± 0.66). No physical activity was associated with a greater A-SC (M = 4.17 ± 0.58; d = 0.697). Adolescents who practiced NCIS (M = 3.13 ± 0.60; d = 1.194) or CTS (M = 2.81 ± 0.57; d = 0.698) had higher values of E-SC than those not practicing any physical activity or sport (M = 2.37 ± 0.67). Individuals who practised NCIS (M = 3.52 ± 0.62; d = 0.922) and CTS (M = 3.49 ± 0.78; d = 0.907) presented the highest values of P-SC in comparison to those who did not practise any activities (M = 2.88 ± 0.76).

Table 5. Practice of physical or sports activities and self-concept during COVID-19 lockdown.

Variable	Sport	Mean	SD	F	Sig.	ES (*d*)	IC 95%
G-SC	NS	3.22	0.36	4.209	$p \leq 0.05$ [a]	0.827 [a]	[0.394; 1.259] [a]
	CIS	3.44	0.12				
	NCIS	3.48	0.26				
	CTS	3.36	0.31				
	NCTS	3.33	0.28				
A-SC	NS	4.17	0.58	3.583	$p \leq 0.05$ [a]	0.697 [a]	[0.269; 1.125] [a]
	CIS	3.66	0.44				
	NCIS	3.64	0.69				
	CTS	3.74	0.67				
	NCTS	3.69	0.78				
S-SC	NS	3.34	0.46	1.531	$p \geq 0.05$	NP	NP
	CIS	3.38	0.22				
	NCIS	3.54	0.35				
	CTS	3.44	0.39				
	NCTS	3.35	0.34				
E-SC	NS	2.37	0.67	1.125	$p \leq 0.05$ [a,b]	1.194 [a] 0.698 [b]	[0.743; 1.645] [a] [0.231; 1.164] [b]
	CIS	2.57	0.58				
	NCIS	3.13	0.60				
	CTS	2.81	0.57				
	NCTS	2.63	0.75				
F-SC	NS	3.51	0.47	0.483	$p \geq 0.05$	NP	NP
	CIS	3.47	0.55				
	NCIS	3.62	0.38				
	CTS	3.49	0.51				
	NCTS	3.53	0.37				
P-SC	NS	2.88	0.76	5.714	$p \leq 0.05$ [a,b]	0.922 [a] 0.907 [b]	[0.485; 1.359] [a] [0.432; 1.382] [b]
	CIS	3.55	0.38				
	NCIS	3.52	0.62				
	CTS	3.49	0.78				
	NCTS	3.33	0.77				

Note 1. General self-concept (G-SC); academic self-concept (A-SC); social self-concept (S-SC); emotional self-concept (E-SC); family self-concept (F-SC); physical self-concept (P-SC). Note 2. No Sport (NS); non-contact individual sport (NCIS); contact individual sport (CIS); non-contact team sport (NCTS); contact team sport (CTS). Note 3. Differences between NP and NCIS ([a]); differences between NP and CTS ([b]).

Figure 5 compares groups of physically active and inactive adolescents during the lockdown. Being physically active during lockdown was associated with higher levels of G-SC, A-SC, S-SC and P-SC. However, no statistically significant differences ($p \geq 0.05$) were found in the E-SC and F-SC dimensions.

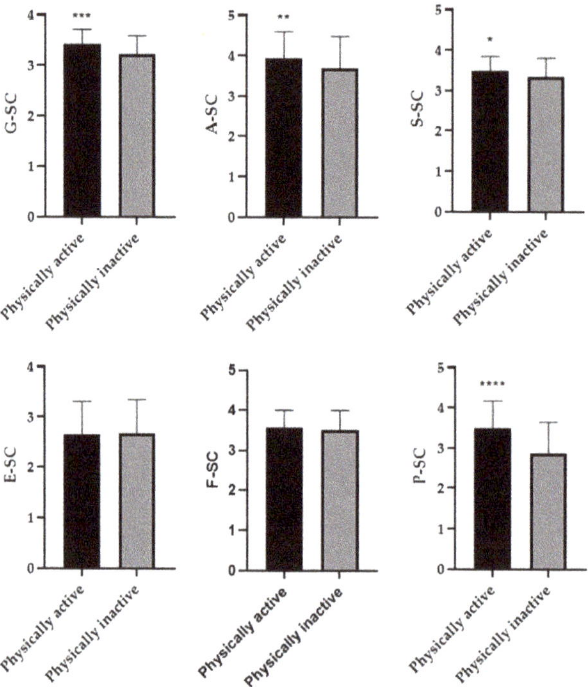

Figure 5. Self-concept and dimensions concerning physically active and inactive adolescents before the lockdown. Note 1. Self-concept (SC); general self-concept (G-SC); academic self-concept (A-SC); social self-concept (S-SC); emotional self-concept (E-SC); family self-concept (F-SC); physical self-concept (P-SC). Note 2. $p \leq 0.05$ (*); $p \leq 0.01$ (**); $p \leq 0.001$ (***); $p \leq 0.0001$ (****).

4. Discussion

This study aims to establish the association betweenself-concept, gender and the practice of physical activities, as well as to estimate the magnitude of their differences, in adolescents before and during the COVID-19 lockdown. There area wide variety of studies that addressed this field before the COVID-19 lockdown [19,37–39], yet evidence evaluating adolescents' self-concept during the COVID-19 pandemic was not found. The strength of this research relies on comparing adolescents' self-concept before and during the COVID-19 pandemic, as well as estimating differences based on gender and the practice of physical activity. This is a crucial area of research since self-concept is a pivotal construct in adolescence [40,41], given its association with quality and life satisfaction, and because it is influenced by social, physical, emotional, academic and family features [40,42,43].

Adolescents' general and social self-concept were higher before the lockdown. However, the emotional detriment during the lockdown caused lower values of self-concept in comparison to before the isolation. In this sense, studies such as Martínez-Marín, Martínez and Paterna [44] evinced that self-concept predicted emotional intelligence, both genders being associated with clarity and emotional repair. In line with this, an appropriate emotional state is associated with a better self-perception, self-esteem and self-rate in youth [45].

Emotional self-concept was higher before the lockdown, and it was correlated to contact with family and peers. Findings reveal that as adolescents' knowledge increases, so positive/negative emotion regulation and psychological adjustment are more efficient [46]. Hence, participants' family self-concept was higher before the lockdown, since adolescents who reported a high family functioning had appropriate values of self-concept [20].

However, during the COVID-19 lockdown there was a negative association between physical and emotional components, as well as higher values of academic self-concept. Staying at home, with resources enough to maintain the normality, meant that adolescents had a higher academic self-concept, which led to higher academic performance [47]. In this sense, perceived competences are demonstrated to buffer the association between academic self-concept and goal-oriented motivation [48].

As regards gender-based differences, females had higher values of academic self-concept in both moments. Moreover, males' academic self-concept was negatively correlated to emotional self-concept, yet family self-concept was positively associated with their academic self-concept during the lockdown. Furthermore, boys had higher levels in physical and emotional self-concept before the lockdown, while girls' emotional self-concept was correlated to support and proximity of family and peers. Males usually have higher emotional and physical self-concept, whereas females tend to have higher academic and family self-concept [19,41,46,48–51]. Age intensifies this difference, and some aspects were revealed, such as peers' relations, academic performance or motor-perceptive skills consolidation [21,52]. On the whole, girls' general self-concept was higher during the lockdown.

The gender-based difference in terms of the dimensions of self-concept causes some differences while choosing a physical activitie, non-contact individual sports and contact team sports being chosen more frequently by adolescents, which are associated with identity acceptance [19,32,53]. However, girls usually are less involved in these types of activity, yet benefits and enhancement on self-concept are very similar if both genders take part in physical activities in the same way [19,54]. Following on from the foregoing, practising no physical activities led to low levels of self-concept before the lockdown; indeed, there was a positive association between self-concept dimensions and the level of physical activity, and self-concept was negatively associated with sedentary behaviour [24]. In line with this, organised physical activity fostered a higher meeting of needs in the youngest [37].

Nevertheless, practising non-contact individual sports was associated with a higher academic self-concept in adolescents, as opposed to those who practised no physical activity. The practice of physical activity and the decrease in sedentary behaviour were generally correlated to academic performance and cognitive development [55–58], so that these factors are positively associated with a higher academic self-concept [59,60]. Likewise, adolescents who practised either non-contact individual sports or contact team sports had higher social, family and physical self-concept in comparison to those who did not practise any physical activity. Generally, it has been demonstrated that the benefits of practising a moderate physical activity in adolescence promote the development of self-concept and help to optimise health and welfare in the long term [61].

During the COVID-19 pandemic, some studies exposed that the practice of physical activity was a useful tool to cope with this situation, as well as to improve people's health and their ability to enhance their self-growth [62,63]. Lack of regular practice of physical activity within this period led to harmful consequences [64–70]. Nevertheless, some countries allowed the off-site practice of physical activity such as biking as a mean of transport, providing that people maintained social distance, which reported a lower health detriment due to both physical activity and social distance [71]. In this investigation, there was a positive association between the practice of individual physical activities and general/physical self-concept in adolescents. Thus, adolescents with higher emotional strength were the ones who expressed practising physical activity. In line with this, psychological consequences during COVID-19 pandemic have also been studied, so the lockdown resulted in harmful psychological consequences, although physical activity was associated with the prevention of psychology disruptions since it helps to control anxiety, decrease psychosomatic symptoms and enhance self-esteem [72–74]. However, adolescents who did not practise any physical activity during the isolation had higher values in academic self-concept. Time expenditure during this lockdown is focused on doing daily tasks such as assignments, practising physical activity and using IT devices [46].

The return to habitual practice of physical activity and its link with COVID-19 is not established yet. Experts are working on protocols that prevent the virus spreading so that people may practise physical or sports activities during a lockdown due to their physical, social, psychological and cognitive benefits [75–78]. This is a crucial matter because this is not the only global pandemic we are supposedly suffering in future. Lastly, it must be highlighted that the World Health Organization [79] states that an active life enhances self-perception [19].

The present study has some limitations. It is important to emphasise some caveats to the findings of this research for generalisation to other populations such as the population size. Furthermore, the descriptive design did not allow for establishing causal association (test-retest), so the results obtained ought to be interpreted cautiously. To the knowledge of the authors of this study, another limitation has been the lack of solid research that worked on the study variables during the COVID-19 pandemic. Consequently, it would be interesting to widen the sample by including more adolescents across the globe so that comparisons may be possible. Moreover, it would be interesting to evaluate the population once the COVID-19 pandemic finishes, establishing associations within the entire period. Lastly, from another perspective and as a practical application, it has been demonstrated that a cognitive-behaviour intervention and the practice of physical activities may help to obtain higher family, emotional and physical self-concept, as well as a decrease in psychopathological symptoms, in harmful situations in which psychosocial problems appear and there is a detriment of mental health [22,30,80,81].

5. Conclusions

The aim of the study was to assess the level of self-concept of adolescents before and during the lockdown, as regards gender and physical activity. Therefore, it is concluded that adolescents' self-rate and self-concept before the lockdown were higher than during the COVID-19 pandemic, which is positively associated with emotional self-concept. Moreover, family and peers were essential factors for an appropriate emotional self-concept development before and during the lockdown. Furthermore, academic self-concept during the lockdown was higher than before it.

It may be highlighted that females' self-concept was higher during the isolation than the males' one. Specifically, girls had a higher academic self-concept in both times, although boys' academic self-concept was influenced by their emotional management and their relationships with family and peers. Even though males' physical and emotional self-concept was higher than the females' ones before the lockdown, these differences did not last during the lockdown.

As regards physical activity, people who did not practise any physical or sports activity had a lower self-concept before the lockdown. During the isolation, practising physical activity before it was associated with higher academic, social, family and physical self-concept, as opposed to those who practised no physical activity. However, academic self-concept was positively associated with not practising physical activity. Within the lockdown, adolescents who practised non-contact individual sports had higher general and emotional self-concept. Lastly, before the lockdown, adolescents who practised physical activity reported higher physical self-concept.

Author Contributions: Conceptualisation, G.G.-V. and F.Z.-O.; methodology, G.G.-V., F.Z.-O. and D.L.-P.; software, G.G.-V. and F.Z.-O.; formal analysis, G.G.-V., F.Z.-O. and J.C.-P.; investigation, G.G.-V., F.Z.-O., D.L.-P. and J.C.-P.; data curation, G.G.-V.; writing—original draft preparation, G.G.-V. and D.L.-P.; writing—review and editing, G.G.-V., F.Z.-O., D.L.-P., G.B. and J.C.-P.; visualisation, G.G.-V., G.B., W.R.G. and F.Z.-O.; supervision, G.G.-V., G.B., W.R.G. and F.Z.-O. All authors have read and agreed to the published version of the manuscript.

Funding: This research received no external funding.

Conflicts of Interest: The authors declare no conflict of interest.

References

1. Baker, L.R.; Cormier, L.A. *Disasters and Vulnerable Populations: Evidence-Based Practice for the Helping Professions*; Springer: New York, NY, USA, 2014.
2. Leser, K.A.; Looper-Coats, J.; Roszak, A.R. Emergency Preparedness Plans and Perceptions Among a Sample of United States Childcare Providers. *Disaster Med. Public Health Prep.* **2019**, *13*, 704–708. [CrossRef]
3. Skowronski, D.M.; Leir, S.; De Serres, G.; Murti, M.; Dickinson, J.A.; Winter, A.-L.; Olsha, R.; Croxen, M.A.; Drews, S.J.; Charest, H.; et al. Children under 10 years of age were more affected by the 2018/19 influenza A(H1N1)pdm09 epidemic in Canada: Possible cohort effect following the 2009 influenza pandemic. *Eurosurveillance* **2019**, *24*, 1900104. [CrossRef]
4. Cinesi, C.; Pañuelas, Ó.; Luján, M.; Egea, C.; Masa, J.F.; García, J.; Ferrando, C. Recomendaciones de consenso respecto al soporte respiratorio no invasivo en el paciente adulto con insuficiencia respiratoria aguda secundaria a infección por SARS-CoV-2. *Med. Intensiva* **2020**, *1480*, 1–10. [CrossRef]
5. World Health Organization. *Coronavirus Disease (COVID-19) Out-Break*; WHO: Geneva, Switzerland, 2020.
6. Zhang, L.; Wang, D.C.; Huang, Q.; Wang, X. Significance of clinical phenomes of patients with COVID-19 infection: A learning from 3795 patients in 80 reports. *Clin. Transl. Med.* **2020**, *10*, 28–35. [CrossRef] [PubMed]
7. Bahn, G.H. Coronavirus Disease 2019, School Closures, and Children's Mental Health. *J. Korean Acad. Child Adolesc. Psychiatry* **2020**, *31*, 74–79. [CrossRef] [PubMed]
8. Szabo, T.G.; Richling, S.; Embry, D.D.; Biglan, A.; Wilson, K.G. From Helpless to Hero: Promoting Values-Based Behavior and Positive Family Interaction in the Midst of COVID-19. *Behav. Anal. Pract.* **2020**, *13*, 568–576. [CrossRef]
9. Liang, L.; Ren, H.; Cao, R.; Hu, Y.; Qin, Z.; Li, C.; Mei, S. The Effect of COVID-19 on Youth Mental Health. *Psychiatr. Q.* **2020**, *91*, 841–852. [CrossRef]
10. Shigemura, J.; Ursano, R.J.; Morganstein, J.C.; Kurosawa, M.; Benedek, D.M. Public responses to the novel 2019 coronavirus (2019-nCoV) in Japan: Mental health consequences and target populations. *Psychiatry Clin. Neurosci.* **2020**, *74*, 281–282. [CrossRef]
11. Cosic, K.; Popovic, S.; Sarlija, M.; Kesedzic, I. Impact of Human Disasters and COVID-19 Pandemic on Mental Health: Potential of Digital Psychiatry. *Psychiatr. Danub.* **2020**, *32*, 25–31. [CrossRef]
12. Brown, M.R.G.; Agyapong, V.I.O.; Greenshaw, A.J.; Cribben, I.; Brett-MacLean, P.; Drolet, J.L.; McDonald-Harker, C.; Omeje, J.; Mankowsi, M.; Noble, S.; et al. After the Fort McMurray wildfire there are significant increases in mental health symptoms in grade 7–12 students compared to controls. *BMC Psychiatry* **2019**, *19*, 18. [CrossRef]
13. Latham, B.S.; Sullivan, J.; Williams, R.S.; Eakin, M.N. Maintaining Emotional Well-Being During the COVID-19 Pandemic: A Resource for Your Patients. *Chronic Obstr. Pulm. Dis. J. COPD Found.* **2020**, *7*, 76–78. [CrossRef]
14. Polizzi, C.; Lynn, S.J.; Perry, A. Stress and coping in the time of covid-19: Pathways to resilience and recovery. *Clin. Neuropsychiatry* **2020**, *17*, 59–62. [CrossRef]
15. Casale, S.; Flett, G.L. Interpersonally-based fears during the COVID-19 pandemic: Reflections on the fear of missing out and the fear of not mattering constructs. *Clin. Neuropsychiatry* **2020**, *17*, 88–93. [CrossRef]
16. Sadhwani, I. Effect of self-concept on adolescent depression. *J. Psychol. Res.* **2013**, *7*, 147–152.
17. Norrington, J. Adolescent Peer Victimization, Self-Concept, and Psychological Distress in Emerging Adulthood. *Youth Soc.* **2020**, 1–23. [CrossRef]
18. Turner, H.A.; Shattuck, A.; Finkelhor, D.; Hamby, S. Effects of Poly-Victimization on Adolescent Social Support, Self-Concept, and Psychological Distress. *J. Interpers. Violence* **2016**, *32*, 755–780. [CrossRef] [PubMed]
19. Alonso, D.A.; Mendo-Lázaro, S.; León-Del-Barco, B.; Mirabel-Alviz, M.; Iglesias-Gallego, D. Multidimensional Self-Concept in Elementary Education: Sport Practice and Gender. *Sustainability* **2018**, *10*, 2805. [CrossRef]
20. Padial-Ruz, R.; Turpin, J.A.P.; Cepero-González, M.; Zurita-Ortega, F. Effects of Physical Self-Concept, Emotional Isolation, and Family Functioning on Attitudes towards Physical Education in Adolescents: Structural Equation Analysis. *Int. J. Environ. Res. Public Health* **2019**, *17*, 94. [CrossRef]
21. Torres, L.H.; Al-Lal, M.; Mohamed, L. Academic Achievement, Self-Concept, Personality and Emotional Intelligence in Primary Education. Analysis by Gender and Cultural Group. *Front. Psychol.* **2020**, *10*, 1–13. [CrossRef]

22. González-Valero, G.; Zurita, F.; Martínez-Martínez, A. Panorama motivacional y de actividad física en estudiantes: Una revisión sistemática. *ESHPA Educ. Sport Health Phys. Act.* **2018**, *1*, 41–58. Available online: http://hdl.handle.net/10481/48961 (accessed on 12 June 2020).
23. Chui, W.H.; Wong, M.Y.H. Gender Differences in Happiness and Life Satisfaction Among Adolescents in Hong Kong: Relationships and Self-Concept. *Soc. Indic. Res.* **2015**, *125*, 1035–1051. [CrossRef]
24. Onetti-Onetti, W.; Chinchilla-Minguet, J.L.; Martins, F.M.L.; Castillo-Rodríguez, A. Self-Concept and Physical Activity: Differences Between High School and University Students in Spain and Portugal. *Front. Psychol.* **2019**, *10*, 1333. [CrossRef] [PubMed]
25. Onyeaka, H.K.; Zahid, S.; Patel, R.S. The Unaddressed Behavioral Health Aspect During the Coronavirus Pandemic. *Cureus* **2020**, *12*, e7351. [CrossRef] [PubMed]
26. Akgül, S.; Hüsnü, Ş.; Derman, O.; Özmert, E.; Bideci, A.; Hasanoğlu, E. Mental health of Syrian refugee adolescents: How far have we come? *Turk. J. Pediatrics* **2019**, *61*, 839–845. [CrossRef]
27. Karatzias, T.; Murphy, P.; Cloitre, M.; Bisson, J.; Roberts, N.; Shevlin, M.; Hyland, P.; Maercker, A.; Ben-Ezra, M.; Coventry, P.; et al. Psychological interventions for ICD-11 complex PTSD symptoms: Systematic review and meta-analysis. *Psychol. Med.* **2019**, *49*, 1761–1775. [CrossRef]
28. Moreno-Alcazar, A.; Treen, D.; Valiente-Gómez, A.; Sio-Eroles, A.; Pérez, V.; Amann, B.L.; Radua, J. Efficacy of Eye Movement Desensitization and Reprocessing in Children and Adolescent with Post-traumatic Stress Disorder: A Meta-Analysis of Randomized Controlled Trials. *Front. Psychol.* **2017**, *8*, 1750. [CrossRef]
29. Villalta, L.; Khadr, S.; Chua, K.-C.; Kramer, T.; Clarke, V.; Viner, R.M.; Stringaris, A.; Smith, P. Complex post-traumatic stress symptoms in female adolescents: The role of emotion dysregulation in impairment and trauma exposure after an acute sexual assault. *Eur. J. Psychotraumatol.* **2020**, *11*, 1710400. [CrossRef]
30. Orrù, G.; Ciacchini, R.; Gemignani, A.; Conversano, C. Psychological intervention measures during the COVID-19 pandemic. *Clin. Neuropsychiatry* **2020**, *17*, 76–79. [CrossRef]
31. Xiang, Y.-T.; Yang, Y.; Li, W.; Zhang, L.; Zhang, Q.; Cheung, T.; Ng, C.H. Timely mental health care for the 2019 novel coronavirus outbreak is urgently needed. *Lancet Psychiatry* **2020**, *7*, 228–229. [CrossRef]
32. Castro-Sánchez, M.; Zurita-Ortega, F.; Martínez-Martínez, A.; Chacón-Cuberos, R.; Espejo-Garcés, T. Motivational climate of adolescents and their relationship to gender, physical activity, sport, federated sport and physical activity family. *RICYDE Rev. Int. Cienc. Deporte.* **2016**, *12*, 262–277. [CrossRef]
33. Shavelson, J.; Hubner, J.J.; Stanton, G.C. Self-concept: Validation of construct interepretations. *Rev. Educ. Res.* **1976**, *46*, 407–442. [CrossRef]
34. García, F.; Musitu, G. *AF5: Self-Concept Form 5*; TEA Ediciones: Madrid, Spain, 1999.
35. Cohen, J. *Statistical Power Analysis for the Behavioral Sciences*, 2nd ed.; Erlbaum: Hillsdale, NJ, USA, 1988.
36. Cohen, J. A power primer. *Psychol. Bull.* **1992**, *112*, 155–159. [CrossRef] [PubMed]
37. Fraguela-Vale, R.; Varela-Garrote, L.; Carretero-García, M.; Peralbo-Rubio, E.M. Basic Psychological Needs, Physical Self-Concept, and Physical Activity Among Adolescents: Autonomy in Focus. *Front. Psychol.* **2020**, *11*, 491. [CrossRef] [PubMed]
38. Garn, A.C.; Morin, A.J.S.; White, R.L.; Owen, K.B.; Donley, W.; Lonsdale, C. Moderate-to-vigorous physical activity as a predictor of changes in physical self-concept in adolescents. *Health Psychol.* **2020**, *39*, 190–198. [CrossRef] [PubMed]
39. Ramírez-Granizo, I.A.; Sánchez-Zafra, M.; Zurita-Ortega, F.; Puertas-Molero, P.; González-Valero, G.; Ubago-Jiménez, J.L. Multidimensional Self-Concept Depending on Levels of Resilience and the Motivational Climate Directed towards Sport in Schoolchildren. *Int. J. Environ. Res. Public Health* **2020**, *17*, 534. [CrossRef] [PubMed]
40. Povedano-Diaz, A.; Povedano-Díaz, A.; Vera, M. Adolescents' Life Satisfaction: The Role of Classroom, Family, Self-Concept and Gender. *Int. J. Environ. Res. Public Health* **2019**, *17*, 19. [CrossRef] [PubMed]
41. Zubeldia, M.; Diaz, M.; Goñi, E. Self-concept, attributions and trait-anxiety in conservatory students. Differences related to age and gender. *Psychol. Soc. Educ.* **2018**, *10*, 79–102. [CrossRef]
42. Ramos-Díaz, E.; Rodríguez-Fernández, A.; Antonio-Agirre, I. Self-concept and subjective well-being being based on gender and educational lever in adolescence. *Psicol. Educ.* **2017**, *23*, 89–94. [CrossRef]
43. Park, S.; Yun, D.-H.; Kwon, I.-S. Relationships of Adolescents' Physical Self-Concept and Life Satisfaction by Gender and Physical Activity Level. *Korean J. Sport Stud.* **2018**, *57*, 195–205. [CrossRef]
44. Martínez-Marín, M.D.; Martínez, C.; Paterna, C. Gendered self-concept and gender as predictors of emotional intelligence: A comparison through of age. *Curr. Psychol.* **2020**, 1–14. [CrossRef]

45. Suriá-Martínez, R.; Ortigosa-Quiles, J.M.; Riquelme-Marin, A. Emotional Intelligence Profiles of University Students with Motor Disabilities: Differential Analysis of Self-Concept Dimensions. *Int. J. Environ. Res. Public Health* **2019**, *16*, 4073. [CrossRef] [PubMed]

46. Parise, M.; Canzi, E.; Olivari, M.G.; Ferrari, L. Self-concept clarity and psychological adjustment in adolescence: The mediating role of emotion regulation. *Pers. Individ. Differ.* **2019**, *138*, 363–365. [CrossRef]

47. Cachón-Zagalaz, J.; Sanabrias-Moreno, D.; Sánchez-Zafra, M.; Lara-Sánchez, A.J.; Lara-Sánchez, A.J. Use of the Smartphone and Self-Concept in University Students According to the Gender Variable. *Int. J. Environ. Res. Public Health* **2020**, *17*, 4184. [CrossRef] [PubMed]

48. Kulakow, S. Academic self-concept and achievement motivation among adolescent students in different learning environments: Does competence-support matter? *Learn. Motiv.* **2020**, *70*, 101632. [CrossRef]

49. Galindo-Domínguez, H. Standardization of the self-concept AF-5 scale by school year and gender in primary education. *Psicol. Educ.* **2019**, *25*, 117–125. [CrossRef]

50. Meza-Pena, C.; Pompa-Guajardo, E.G. Gender, obesity and self concept in a sample of Mexican adolescents. *RICYDE Rev. Int. Cien. Dep.* **2016**, *12*, 137–148. [CrossRef]

51. Tapia-López, A. Gender differences in physical activity levels, degree of adherence to the Mediterranean diet, and physical self-concept in adolescents. *Retos* **2019**, *36*, 185–192.

52. Xie, F.; Xin, Z.; Chen, X.; Zhang, L. Gender Difference of Chinese High School Students' Math Anxiety: The Effects of Self-Esteem, Test Anxiety and General Anxiety. *Sex Roles* **2018**, *81*, 235–244. [CrossRef]

53. Brown, D.M.; Cairney, J. The synergistic effect of poor motor coordination, gender and age on self-concept in children: A longitudinal analysis. *Res. Dev. Disabil.* **2020**, *98*, 103576. [CrossRef]

54. Chalabaev, A.; Sarrazinr, P.; Fontayne, P.; Boiché, J.; Clément-Guillotin, C. The influence of sex stereotypes and gender roles on participation and performance in sport and exercise: Review and future directions. *Psychol. Sport Exerc.* **2013**, *14*, 136–144. [CrossRef]

55. Clevinger, K.; Petrie, T.; Martin, S.; Greenleaf, C. The Relationship of Sport Involvement and Gender to Physical Fitness, Self-Efficacy, and Self-Concept in Middle School Students. *Phys. Educ.* **2020**, *77*, 154–172. [CrossRef]

56. Gómez-Fernández, N.; Albert, J.-F. Physical activity in and out-of-school and academic performance in Spain. *Health Educ. J.* **2020**, *79*, 1–15. [CrossRef]

57. Masoomi, H.; Taheri, M.; Irandoust, K.; H'Mida, C.; Chtourou, H. The relationship of breakfast and snack foods with cognitive and academic performance and physical activity levels of adolescent students. *Biol. Rhythm. Res.* **2019**, *51*, 481–488. [CrossRef]

58. Ishii, K.; Aoyagi, K.; Shibata, A.; Koohsari, M.J.; Carver, A.; Oka, K. Joint Associations of Leisure Screen Time and Physical Activity with Academic Performance in a Sample of Japanese Children. *Int. J. Environ. Res. Public Health* **2020**, *17*, 757. [CrossRef]

59. Carriedo, A.; González, C. Academic Achievement in Physical Education: Academic versus Physical Activity aspects. *Cult. Cienc. Deporte* **2019**, *14*, 225–232. [CrossRef]

60. Louise, T.; Hernández, A.; Reigal, R.E.; Morales, V. Effects of physical activity on self-concept and self-efficacy in preadolescents. *Retos* **2016**, *29*, 61–65.

61. Grant, P.M.; Perivoliotis, D.; Luther, L.; Bredemeier, K.; Beck, A.T. Rapid improvement in beliefs, mood, and performance following an experimental success experience in an analogue test of recovery-oriented cognitive therapy. *Psychol. Med.* **2017**, *48*, 261–268. [CrossRef]

62. Neto, L.D.O.; Elsangedy, H.M.; Tavares, V.D.D.O.; Teixeira, C.V.L.S.; Behm, D.G.; Da Silva-Grigoletto, M.E. TrainingInHome—Home-based training during COVID-19 (SARS-COV2) pandemic: Physical exercise and behavior-based approach. *Rev. Bras. Fisiol. Exerc.* **2020**, *19*, 9–19. [CrossRef]

63. Aubertin-Leheudre, M.; Rolland, Y. The Importance of Physical Activity to Care for Frail Older Adults during the COVID-19 Pandemic. *J. Am. Med. Dir. Assoc.* **2020**, *21*, 973–976. [CrossRef]

64. Vetrovsky, T.; Frybova, T.; Gant, I.; Semerad, M.; Cimler, R.; Bunc, V.; Siranec, M.; Miklikova, M.; Vesely, J.; Griva, M.; et al. The detrimental effect of COVID-19 nationwide quarantine on accelerometer-assessed physical activity of heart failure patients. *ESC Hear. Fail.* **2020**, 1–5. [CrossRef]

65. Qin, F.; Song, Y.; Nassis, G.P.; Zhao, L.; Cui, S.; Lai, L.; Wu, Z.; Xu, M.; Qu, C.; Dong, Y.; et al. Prevalence of Insufficient Physical Activity, Sedentary Screen Time and Emotional Well-Being During the Early Days of the 2019 Novel Coronavirus (COVID-19) Outbreak in China: A National Cross-Sectional Study. *SSRN Electron. J.* **2020**, 1–28. [CrossRef]

66. Chen, P.; Mao, L.; Nassis, G.P.; Harmer, P.; Ainsworth, B.E.; Li, F. Coronavirus disease (COVID-19): The need to maintain regular physical activity while taking precautions. *J. Sport Health Sci.* **2020**, *9*, 103–104. [CrossRef] [PubMed]

67. Lippi, G.; Henry, B.M.; Sanchis-Gomar, F. Physical inactivity and cardiovascular disease at the time of coronavirus disease 2019 (COVID-19). *Eur. J. Prev. Cardiol.* **2020**, *27*, 906–908. [CrossRef] [PubMed]

68. Di Stefano, V.; Battaglia, G.; Giustino, V.; Gagliardo, A.; D'Aleo, M.; Giannini, O.; Palma, A.; Brighina, F. Significant reduction of physical activity in patients with neuromuscular disease during COVID-19 pandemic: The long-term consequences of quarantine. *J. Neurol.* **2020**, 1–7. [CrossRef]

69. Van De Venis, L.; Van De Warrenburg, B.P.C.; Weerdesteyn, V.; Van Lith, B.J.H.; Geurts, A.C.H.; Nonnekes, J. COVID-19 reveals influence of physical activity on symptom severity in hereditary spastic paraplegia. *J. Neurol.* **2020**, *236*, 1–3. [CrossRef]

70. Lim, M.A.; Pranata, R. Sports activities during any pandemic lockdown. *Ir. J. Med Sci.* **2020**, 1–5. [CrossRef]

71. Park, S.; Kim, B.; Lee, J. Social Distancing and Outdoor Physical Activity During the COVID-19 Outbreak in South Korea: Implications for Physical Distancing Strategies. *Asia Pac. J. Public Health* **2020**, *32667221*, 1–3. [CrossRef]

72. Dwyer, M.J.; Pasini, M.; De Dominicis, S.; Righi, E. Physical activity: Benefits and challenges during the COVID-19 pandemic. *Scand. J. Med. Sci. Sports* **2020**, *30*, 1291–1294. [CrossRef]

73. Pieh, C.; Budimir, S.; Probst, T. The effect of age, gender, income, work, and physical activity on mental health during coronavirus disease (COVID-19) lockdown in Austria. *J. Psychosom. Res.* **2020**, *136*, 110186. [CrossRef]

74. Shahidi, S.H.; Williams, J.S.; Hassani, F. Physical activity during COVID-19 quarantine. *Acta Paediatr.* **2020**, 1–2. [CrossRef]

75. Blocken, B.; Van Druenen, T.; Van Hooff, T.; Verstappen, P.; Marchal, T.; Marr, L. Can indoor sports centers be allowed to re-open during the COVID-19 pandemic based on a certificate of equivalence? *Build. Environ.* **2020**, *180*, 107022. [CrossRef] [PubMed]

76. Buldú, J.; Antequera, D.R.; Aguirre, J. The resumption of sports competitions after COVID-19 lockdown: The case of the Spanish football league. *Chaos Solitons Fractals* **2020**, *138*, 109964. [CrossRef] [PubMed]

77. Stokes, K.A.; Jones, B.; Bennett, M.; Close, G.L.; Gill, N.; Hull, J.H.; Kasper, A.M.; Kemp, S.P.; Mellalieu, S.D.; Peirce, N.; et al. Returning to Play after Prolonged Training Restrictions in Professional Collision Sports. *Int. J. Sports Med.* **2020**, 2–3. [CrossRef] [PubMed]

78. Matias, T.S.; Dominski, F.H.; Marks, D. Human needs in COVID-19 isolation. *J. Health Psychol.* **2020**, *25*, 871–882. [CrossRef] [PubMed]

79. World Health Organization (WHO). *Global Recommendations on Physical Activity for Health*; WHO Press: Geneva, Switzerland, 2010; pp. 15–21.

80. Miotto, K.; Sanford, J.; Brymer, M.J.; Bursch, B.; Pynoos, R.S. Implementing an emotional support and mental health response plan for healthcare workers during the COVID-19 pandemic. *Psychol. Trauma Theory Res. Pract. Policy* **2020**, *12*, S165–S167. [CrossRef]

81. Ramos-Díaz, E.; Axpe, I.; Fernández-Lasarte, O.; Jiménez-Jiménez, V. Cognitive-behavioral intervention to enhance self-concept in a case of an adolescent victim of emotional maltreatment. *Rev. Clin. Contemp.* **2018**, *9*, 1–10. [CrossRef]

 *sustainability*

Article

Optimal Body Composition and Anthropometric Profile of World-Class Beach Handball Players by Playing Positions

Basilio Pueo *, Jose Julio Espina-Agullo, Sergio Selles-Perez and Alfonso Penichet-Tomas

Department of General and Specific Didactics, University of Alicante, 03690 Alicante, Spain; jj.espina@ua.es (J.J.E.-A.); sergio.selles@ua.es (S.S.-P.); alfonso.penichet@ua.es (A.P.-T.)

* Correspondence: basilio@ua.es

Received: 1 August 2020; Accepted: 18 August 2020; Published: 21 August 2020

Abstract: Profiling of beach handball players is required to optimize sports performance, talent identification, and injury prevention. The study aimed to describe the anthropometric characteristics, somatotype, and body composition of elite male and female beach handball players classified by playing positions. Thirty elite beach handball players (15 male, 15 female) of the Spanish National Beach Handball Team, which ranked fifth and first in the VII World Championships, respectively, were categorized as front (wings/specialists), back (pivots/defenders) players and goalkeepers. Independent from position, male players showed larger values of anthropometric characteristics, girths, breadths, and absolute components of body composition than female players. Contrastingly, skinfolds, and body fat mass percentage were higher in female players. All these results were statistically significant ($p < 0.05$) with large to extremely large effect sizes ($d = 1.4$–5.4). The position-related differences indicated that male back players were taller ($p = 0.008$; $\eta_p^2 = 0.56$), heavier ($p = 0.016$; $\eta_p^2 = 0.50$) and showed larger arm span ($p = 0.036$; $\eta_p^2 = 0.42$) than front players. In contrast, female goalkeepers showed larger body mass ($p = 0.007$; $\eta_p^2 = 0.57$) and BMI ($p = 0.035$; $\eta_p^2 = 0.43$), whereas back players showed higher muscular mass than goalkeepers ($p = 0.022$; $\eta_p^2 = 0.47$). The present study provides anthropometric reference values of elite beach handball players, and indicates differences between playing positions, providing normative data for talent identification of male and female players.

Keywords: team sports; fat mass; muscle mass; elite players; positional differences; sand; talent identification

---

1. Introduction

Beach handball emerged as a sport derived from team handball with distinctive rules and a sandy playing surface. In recent years, beach handball has become increasingly popular thanks to the support of various bodies, such as the International Handball Federation and the International Olympic Committee. As a result of this popularity, the first International Handball Federation (IHF) Beach Handball World Championships was organized in 2004 on a biannual basis. Since then, the IHF has also organized indoor youth, junior and senior World Championships, as well as Asian, European, Oceania, and Pan American Championships. Additionally, beach handball is played in national leagues from more than 50 countries. In 2018, the last world tournament was the Beach Handball World Championships (Kazan, Russia). From 30 June to 5 July 2020, the Beach Handball World Championships has recently taken place in Pescara, Italy, with 400 athletes from 32 teams and 22 nations from the five continents. As beach handball continues to grow in popularity, it is now being considered to debut as a separate event in the 2024 Olympic Games, as a step towards becoming an Olympic sport. For the above reasons, beach handball can be regarded as a sport with a huge impact among coaches, participants, and spectators.

Beach handball and team handball share motor characteristics such as accelerations, sprints or jumps, as well as rapid changes of direction and a high number of physical collisions [1–5]. The intense and intermittent nature of the two variants of the sport suggests that handball players must be able to develop high values of maximum strength and muscle power in both upper and lower limbs [6], so muscle mass, and therefore body composition, are factors in fitness and success [7–10]. Along with body mass, other anthropometric characteristics have been shown crucial for sporting action and performance in handball [11–13] and other typical team sports such as volleyball, soccer and rugby [14]. For instance, throwing, as the most important technical action in handball players performance [1,15,16], depends on the ability of the arm to reach sufficient acceleration so that the ball leaves the hand at maximum possible speed. The duration of the throwing movement reduces the visual information available to the goalkeeper and the speed of the ball is related to the time for the goalkeeper to save a goal [17]. As a result, bigger body size in terms of fat-free mass and pertinent anthropometrics have positive effects on throwing performance. For instance, increased hand spread helps firm ball grip [13,16]. Besides, the knowledge of anthropometric body measures is of paramount importance for successful talent identification programs as it fulfills four of the five requirements identified by Ackland [18]: (a) recording a set of data of each athlete, (b) gathering a set of normative data, (c) using these data to construct profiles of athletes, and (d) interpreting such profiles to guide the selection process or provide the basis for an ongoing training program. Therefore, anthropometric profile and somatotype give valuable information about handball players' physical condition and allow coaches to identify talent, select players and provide appropriate training volumes and intensities to increase their capabilities.

The anthropometric profile of team handball players has been widely reported in the literature [3,15,19–24], where reference values were used for player identification and selection criteria. However, despite the growth of beach handball at a participatory and organizational level in recent years, the only studies addressing the anthropometric profile of beach handball players have focused on elite female players with no indication of playing position [10,25,26]. Therefore, there is no information available about the variation of anthropometric characteristics by gender and playing positions in elite beach handball players. The comparison between male and female players would provide evidence of the differences between beach handball teams, especially for top-level elite players. Despite variations between genders are expected, to the knowledge of the authors, there is no literature addressing them in a beach handball elite sample. In the field of sports anthropometry, the comparison between male and female helps to understand the variation in specific characteristics for elite [27] and non-elite players [28]. Similarly, to our best knowledge, there are no studies indicating the differences and the distinctive pattern between playing positions in beach handball. Thus, there is not enough evidence of anthropometric characteristics by playing positions in this discipline as with team handball [29,30]. An anthropometric analysis of male and female players' profile would help in identifying talent and optimizing the strength and conditioning training programs for each playing position [2,22,31].

Therefore, this study aims to provide anthropometric reference framework for elite beach handball players and explore how these parameters differ between gender and between playing positions. To this effect, we carried out a comparative analysis of the anthropometric profile, somatotype and body composition by playing position of male and female elite beach handball players of the Spanish national team. This quality sample comprises the entire Spanish National Beach Handball Team competing at international level in World Championships.

2. Materials and Methods

2.1. Subjects

The study sample was composed of 15 male and 15 female elite beach handball players participating in the Annual Spanish Beach Handball Cup. They all were professional players belonging to the

National Beach Handball Team of the Royal Spanish Handball Federation which ranked fifth (male) and first (female) in the World Championships. This sample represented the population of Spanish male and female international elite players. Players were categorized as front players (wings and specialists), back players (pivots and defenders) and goalkeepers, according to position-specific playing demands. Subjects were instructed to conduct normal dietary habits and report to the measurement tent in a fully hydrated state. All participants were previously informed about the objectives of the research, the experimental protocol and the procedures of the study and voluntarily gave written informed consent to participate in the study in accordance with the Declaration of Helsinki.

2.2. Anthropometric Data

Anthropometric measurements were performed following standard protocols adopted by the International Society for the Advancement of Kinanthropometry (ISAK) [32]. All measurements were taken in basal conditions, in the same tent, at ambient temperature (22 ± 1 °C), and by the same researcher who was an accredited Level 2 anthropometrist of ISAK. Technical measurement error was lower than 5% for skinfolds and lower than 1% for girths and breadths.

Seventeen anthropometric variables were measured for each subject. Height and body mass were measured on portable set scales (models 213 and 707, Seca, Hamburg, Deutschland) to the nearest 0.1 cm and 0.01 kg, respectively. The thicknesses of 8 skinfolds (subscapular, triceps, biceps, iliac crest, supraspinale, abdominal, front thigh and medial calf) were measured using a caliper calibrated to the nearest 0.2 mm (Holtain Ltd., Crymych, UK). The sum of 6 skinfolds was also computed (subscapular, triceps, supraspinale, abdominal, front thigh and medial calf). Four girths (relaxed arm, flexed arm, thigh and calf), and 3 breadths (humerus, stylion and femur) measurements were performed using a flexible anthropometric steel tape (Holtain Ltd., Crymych, UK) to the nearest 0.1 cm. Body composition was calculated using the following models: fat mass was computed through the methods of Withers, Craig, Bourdon, and Norton [33], muscle and bone masses were determined using the methods of Lee et al. [34] and Rocha [35], respectively. According to the Kinanthropometry Spanish Committee, these methods are the most appropriate for high-performance players [36]. Mean somatotype was determined using the Heath and Carter anthropometric method [37] and its classification according to the categories by Carter and Heath [38].

2.3. Statistical Analyses

Standard descriptive statistics was used to show participant characteristics for all variables (Mean ± SD). The Kolmogorov–Smirnov and Levene tests were applied to check sample normality. Independent samples *t*-test was used to compare anthropometric data between male and female groups with statistical significance set at $p < 0.05$. The Cohen's d was used as a measure of the effect size of differences between male and female players and interpreted according to Cohen's thresholds [39] modified by Hopkins [40] as small ($d = 0.2$), moderate ($d = 0.6$), large ($d = 1.2$), very large ($d = 2.0$) and extremely large ($d = 4.0$). Mean differences of selected anthropometric characteristics, body composition and somatotype components of male and female players between playing positions were tested using a one-way univariate general linear model with a Tukey post hoc test ($p < 0.05$). As in a similar study on elite team handball players [30], the decision of significance was based on eta-squared $\eta_p^2 > 0.2$ to avoid overestimation of mean differences given the small frequency of position-related cases. The Somatotype Attitudinal Mean (SAM) and the Somatotype Attitudinal Variance (SAV) was used to describe the magnitude of the absolute scatter of the group of somatotypes around each group mean for both male and female players, and for positions within each group. Likewise, the Somatotype Attitudinal Distance (SAD) was used to compare somatotype group means of male and female players, and between positions. All statistical analyses were performed using IBM SPSS Statistics for Windows, version 22 (IBM Corp, Armonk, NY, USA).

3. Results

Table 1 shows the basic anthropometric and demographic characteristics of the sample and the results from the independent samples t-test between male and female players. Mean height and body mass were 187.4 ± 8.2 cm and 85.2 ± 11.3 kg for male players, and 169.1 ± 5.1 cm and 62.9 ± 5.3 kg for female players. Except for age, male players showed larger values in all characteristics than female players. These differences were statistically significant with moderate to very large effect size values.

Table 1. Anthropometric and demographic characteristics of the sample.

Variable	Male		Female		Ind. Samples *t*-Test	
	Mean ± SD	Range	Mean ± SD	Range	*p*	Cohen's *d*
Age (years)	27.1 ± 5.2	20.0–37.0	24.1 ± 4.7	17.0–33.0	0.10	0.6 (moderate)
Body height (cm)	187.4 ± 8.2 *	169.5–202.6	169.1 ± 5.1	160.0–177.9	<0.001	2.7 (very large)
Body mass (kg)	85.2 ± 11.3 *	66.6–112.9	62.9 ± 5.3	55.8–72.7	<0.001	2.5 (very large)
Arm span (cm)	192.6 ± 12.3 *	174.5–228.0	170.6 ± 4.7	164.4–179.4	<0.001	2.4 (very large)
BMI (kg/m^2)	24.2 ± 2.5 *	19.5–28.8	22.0 ± 1.5	19.8–24.5	<0.05	1.1 (moderate)
FMI (kg/m^2)	2.9 ± 1.2	1.4–6.2	3.4 ± 1.0	2.0–5.7	0.18	0.5 (small)

* Statistical significance between male and female players; BMI: Body Mass Index, FMI: Fat Mass Index, Cohen's *d* (Effect Size).

The three groups of anthropometric characteristics shown in Table 2 depict a general tendency for female players to have larger values of skinfolds and male players to show larger girths and breadths values, most of them statistically significant at $p < 0.01$ with moderate to very large effect size values.

Table 2. Descriptive statistics for skinfolds, girths, lengths and breadths and difference between male and female players.

	Variable	Male		Female		Ind. Samples *t*-Test	
		Mean ± SD	Range	Mean ± SD	Range	*p*	Cohen's *d*
Skinfolds	Subscapular (mm)	10.9 ± 4.6	6.8–25.6	11.0 ± 6.3	7.2–31.0	0.95	0.02 (trivial)
	Triceps (mm)	7.9 ± 3.1	3.8–15.4	12.2 ± 4.0 *	5.2–21.2	0.03	1.2 (large)
	Biceps (mm)	3.5 ± 0.7	2.8–5.2	6.1 ± 2.7 *	3.4–11.6	<0.01	1.3 (large)
	Iliac crest (mm)	13.4 ± 6.4	6.2–27.5	17.2 ± 5.2	9.4–25.8	0.08	0.7 (moderate)
	Supraspinale (mm)	10.1 ± 7.8	4.0–36.2	11.0 ± 3.5	6.4–18.2	0.71	0.1 (trivial)
	Abdominal (mm)	15.1 ± 9.1	3.8–37.2	16.0 ± 4.8	8.4–26.2	0.76	0.1 (trivial)
	Front thigh (mm)	13.2 ± 5.3	5.4–21.2	26.1 ± 8.0 *	10.2–37.6	<0.001	1.9 (large)
	Medial calf (mm)	5.7 ± 2.1	3.6–10.4	13.1 ± 5.6 *	4.8–23.0	<0.001	1.8 (large)
	Σ 6 skinfolds (mm)	62.9 ± 24.1	32.4–125.4	89.4 ± 24.2 *	49.3–138.8	<0.01	1.1 (moderate)
Girths	Relaxed arm (cm)	31.8 ± 3.2 *	24.4–36.2	25.1 ± 2.7	22.1–33.0	<0.001	2.3 (very large)
	Flexed arm (cm)	34.2 ± 2.9 *	27.0–37.6	26.5 ± 2.5	23.2–33.4	<0.001	2.8 (very large)
	Thigh (cm)	49.0 ± 12.6	40.6–57.7	45.3 ± 4.5	40.6–60.0	0.29	0.4 (small)
	Calf (cm)	36.7 ± 2.7 *	32.4–41.1	32.4 ± 2.0	29.0–37.1	<0.001	1.8 (large)
Breadths	Humerus (cm)	7.3 ± 0.6 *	6.2–8.2	6.4 ± 0.3	5.9–6.9	<0.001	1.9 (large)
	Stylion (cm)	5.9 ± 0.7 *	5.1–7.7	5.1 ± 0.3	4.7–5.8	<0.001	1.5 (large)
	Femur (cm)	9.7 ± 1.1 *	6.2–10.5	9.0 ± 0.5	8.3–9.9	0.03	0.8 (moderate)

* Statistical significance between male and female players; BMI: Body Mass Index, Cohen's *d* (Effect Size).

Table 3 shows somatotype differences between male and female players in endomorphy and mesomorphy, with statistical significance and large effect size values. As shown in Figure 1, the mean somatotype could be defined as balanced mesomorph (2.6-4.4-2.7) with SAM = 1.85 and endomorph-mesomorph (3.5-3.3-2.6) with SAM = 1.35 for male and female players, respectively. The difference between the two groups' somatotypes SAD is 1.46.

Table 3. Somatotype components and difference between male and female players.

Variable	Male		Female		Ind. Samples *t*-Test	
	Mean ± SD	Range	Mean ± SD	Range	*p*	Cohen's *d*
Endomorphy	2.6 ± 1.1	1.3–5.7	3.5 ± 1.0 *	2.4–5.8	0.03	0.8 (moderate)
Mesomorphy	4.4 ± 1.4 *	1.8–6.4	3.3 ± 0.7	1.9–4.9	<0.01	1.0 (moderate)
Ectomorphy	2.7 ± 1.3	0.6–5.7	2.6 ± 0.8	1.2–6.4	0.81	0.1 (trivial)
SAM	1.85 ± 1.09	0.17–3.38	1.35 ± 0.57	0.45–2.62	0.12	0.6 (moderate)
Ponderal index	42.7 ± 1.8	39.9–46.8	42.6 ± 1.1	40.6–44.5	0.81	0.1 (trivial)

* Statistical significance between male and female players; BMI: Body Mass Index, Cohen's *d* (Effect Size).

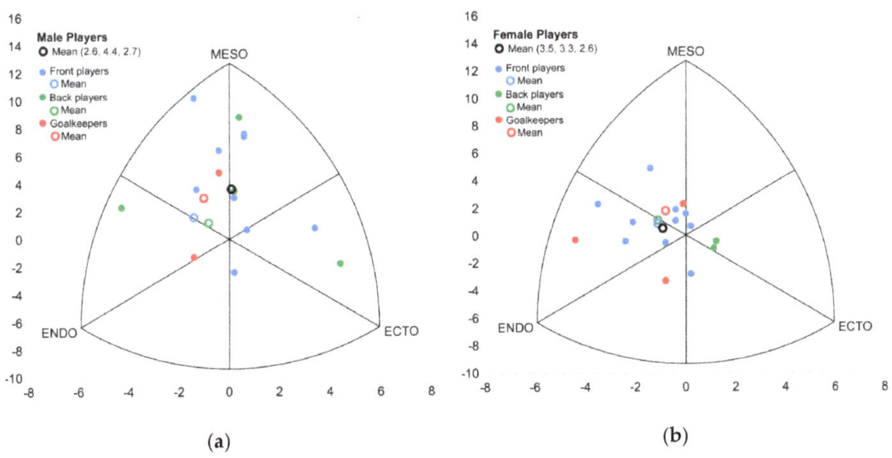

Figure 1. Somatocharts of beach handball players showing plots of the study sample by position and mean somatotypes (endomorph, mesomorph, ectomorph). (**a**) Male players, (**b**) Female players.

Table 4 shows that male players had greater muscular ($p < 0.01$), bone ($p < 0.1$) and residual masses ($p < 0.1$) than female players with extremely to large effect size values. Female players were characterized by larger values of body fat, although not statistically significant. The percentage values showed a similar trend to the absolute values.

Table 4. Body composition and difference between male and female players.

Variable	Male		Female		Ind. Samples *t*-Test	
	Mean ± SD	Range	Mean ± SD	Range	*p*	Cohen's *d*
Muscular mass (kg)	36.2 ± 3.1 *	31.5–43.6	22.9 ± 1.5	20.6–24.9	<0.001	5.4 (ex. large)
Muscular mass (%)	42.7 ± 2.6 *	38.7–47.4	36.4 ± 1.2	34.3–38.5	<0.001	3.0 (very large)
Body fat mass (kg)	10.2 ± 4.8	5.6–24.3	9.8 ± 2.9	5.6–16.0	0.76	0.1 (trivial)
Body fat mass (%)	11.7 ± 3.9	6.9–21.5	15.4 ± 3.7 *	10.1–23.2	0.01	0.9 (moderate)
Bone mass (kg)	13.3 ± 1.8 *	10.8–16.7	9.9 ± 0.9	8.3–11.0	<0.01	2.4 (very large)
Bone mass (%)	15.7 ± 1.6	13.5–20.3	15.7 ± 1.1	13.8–17.4	0.98	0.0 (trivial)
Residual mass (kg)	25.4 ± 4.4 *	17.0–33.0	20.4 ± 2.6	17.2–26.3	<0.01	1.4 (large)
Residual mass (%)	29.8 ± 3.5	25.0–34.4	32.4 ± 3.4	27.7–40.7	0.05	0.7 (moderate)

* Statistical significance between male and female players; BMI: Body Mass Index, Cohen's *d* (Effect Size).

Results in Table 5 also indicate differences in male players' position for body height ($p = 0.008$; $\eta_p^2 = 0.56$), body mass ($p = 0.016$; $\eta_p^2 = 0.50$) and arm span ($p = 0.036$; $\eta_p^2 = 0.42$). Post hoc analyses revealed that back players were taller (+12.7 cm, $p = 0.006$), heavier (+16.3 kg, $p = 0.014$) and showed larger arm span (+ 16.6 cm, $p = 0.029$) than front players. In contrast, position-related differences in female players were seen for body mass ($p = 0.007$; $\eta_p^2 = 0.57$), BMI ($p = 0.035$; $\eta_p^2 = 0.43$) and muscular mass ($p = 0.022$; $\eta_p^2 = 0.47$). Goalkeepers were the female players with the highest body mass (+ 8.8 kg,

$p = 0.010$ vs. front players; +11.7 kg, $p = 0.013$ vs. back players) and BMI (+3.4 kg/m^2, $p = 0.029$ vs. back players), whereas female back players showed higher muscular mass than goalkeepers (+2.7%, $p = 0.021$).

Table 5. Position-related differences in selected anthropometric characteristics, body composition and somatotype components of male and female players (Mean ± SD).

	Variable	Front Players	Back Players	Goalkeepers	ANOVA		
					F	p	η_p^2
Male players	Body height (cm)	181.4 ± 6.4	194.1 ± 5.7 #	187.9 ± 2.0	7.52	0.008	0.56
	Body mass (kg)	77.1 ± 7.2	93.4 ± 10.7 #	88.6 ± 0.8	5.92	0.016	0.50
	Arm span (cm)	185.2 ± 6.2	201.8 ± 13.9 #	191.1 ± 3.7	4.43	0.036	0.42
	BMI (kg/m^2)	23.5 ± 2.4	24.8 ± 2.9	25.0 ± 0.8	0.56	0.586	0.08
	FMI (kg/m^2)	2.4 ± 0.6	3.3 ± 1.6	3.3 ± 1.0	1.02	0.391	0.14
	Σ 6 skinfolds (mm)	54.9 ± 12.4	69.2 ± 32.7	72.3 ± 31.5	0.71	0.511	0.11
	Endomorphy	2.1 ± 0.6	2.9 ± 1.6	4.1 ± 1.7	1.02	0.391	0.14
	Mesomorphy	5.0 ± 1.1	4.0 ± 1.6	3.6 ± 1.4	1.21	0.331	0.17
	Ectomorphy	2.7 ± 1.3	2.8 ± 1.5	2.3 ± 0.4	0.11	0.894	0.02
	Muscular mass (%)	43.9 ± 2.5	41.8 ± 2.9	41.6 ± 0.7	1.33	0.300	0.18
	Body fat mass (%)	10.3 ± 2.3	12.8 ± 5.1	13.2 ± 4.4	0.84	0.454	0.12
Female players	Body height (cm)	167.1 ± 4.1	171.7 ± 5.4	174.0 ± 5.1	3.24	0.075	0.35
	Body mass (kg)	61.5 ± 4.0	58.5 ± 3.9	70.3 ± 2.1 § †	7.84	0.007	0.57
	Arm span (cm)	169.4 ± 3.9	170.5 ± 6.7	174.6 ± 5.5	1.59	0.244	0.21
	BMI (kg/m^2)	22.0 ± 1.3	19.8 ± 0.1	23.3 ± 1.4 †	4.47	0.035	0.43
	FMI (kg/m^2)	3.3 ± 0.8	2.4 ± 0.5	4.5 ± 1.1	3.71	0.056	0.38
	Σ 6 skinfolds (mm)	87.6 ± 20.7	68.8 ± 18.7	108.9 ± 31.4	1.97	0.182	0.25
	Endomorphy	3.5 ± 0.8	2.6 ± 0.3	4.1 ± 1.7	1.31	0.306	0.18
	Mesomorphy	3.4 ± 0.8	2.9 ± 0.1	3.0 ± 0.7	0.76	0.487	0.11
	Ectomorphy	2.4 ± 0.7	3.8 ± 0.3	2.3 ± 0.9	3.25	0.075	0.35
	Muscular mass (%)	36.69 ± 1.1	37.8 ± 0.1 †	35.0 ± 0.6	5.35	0.022	0.47
	Body fat mass (%)	15.0 ± 3.0	12.0 ± 2.7	19.1 ± 4.3	3.15	0.079	0.34

Statistical significance between Front and back players (#); Front players and Goalkeepers (§); Back players and Goalkeepers (†); η_p^2: ANOVA's eta-squared.

As shown in Figure 1, the position-related male somatotypes resulted in mesomorph-endomorph, 3.6-3.7-2.2 for front players, (SAM = 1.7); mesomorph-endomorph, 3.3-3.5-2.5 for back players, (SAM = 2.2) and endomorphic mesomorph, 3.0-4.0-2.0 for goalkeepers, (SAM = 1.5). The differences between mean somatotypes were: front vs. back players, SAD = 0.4; front players vs. goalkeepers, SAD = 0.7; and back players vs. goalkeepers, SAD = 0.7. Besides, female players showed a mesomorph-endomorph profile for front players 3.5-3.4-2.4 (SAM = 1.2); mesomorph-endomorph for back players 3.4-3.4-2.3 (SAM = 1.6); and endomorphic mesomorph for goalkeepers 3.3-3.8-2.5 (SAM = 1.8) with differences between front and back players, SAD = 0.2; front players vs. goalkeepers, SAD = 0.4; and back players vs. goalkeepers, SAD = 0.4.

4. Discussion

The assessment of anthropometric characteristics, somatotype and body composition in beach handball players can be considered a challenging area of study due to the limited population of athletes in elite category. To the best knowledge of the authors, this is the first study providing an anthropometric reference framework for beach handball elite players and differences in body measures as a function of gender and playing positions. The strength of the present study is the high quality of the study sample, all participants being elite players from the Spanish National Selection competing at international level.

A comparison of the main characteristics: age, height, body mass, BMI, body fat mass percentage and somatotype of beach and team handball players is shown in Table 6. Anthropometric characteristics of the study sample show that male elite beach handball players show larger values of body height (187.4 ± 8.2 cm) and body mass (85.2 ± 11.3 kg) than female players (169.1 ± 5.1 cm and 62.9 ± 5.3 kg, respectively). These latter values are in accordance with those previously reported in two studies on beach handball female players from Spain: 167.87 ± 4.42 cm; 61.04 ± 3.98 kg, [25], 168.00 ± 3.86 cm;

60.78 ± 3.87 kg [5] and in another two similar studies from Brazil: 169.50 ± 8.09 cm; 65.43 ± 9.44 kg [10], 168.0 ± 10.0 cm; 63.8 ± 7.1 kg [26]. Likewise, in the study of Zapardiel and Asín-Izquierdo [5], male players showed similar values of body height (187.52 ± 7.48 cm) and body mass (86.96 ± 9.53 kg) to our sample. These results are also in accordance with other team sports such as basketball [41,42], volleyball [43,44] or football [42,45], in which male players reported higher height, body mass and BMI values than female players.

According to our results, male beach handball players are shorter and lighter than their team counterparts from Spain (192.88 ± 7.60 cm and 96.88 ± 11.23 kg), and also than all other nationalities (190.10 ± 6.82 cm and 92.37 ± 9.80 kg) [46]. To the best knowledge of the authors, no similar studies have addressed the anthropometric characteristics, somatotype and body composition in beach handball players by playing position. In the field of team handball, male beach handball goalkeepers in our study show similar basic anthropometric characteristics (height, body mass and BMI than a broad sample of players participating in a team Handball World Championships (191.89 ± 5.18 cm; 95.6 ± 10.45 kg; 25.97 ± 2.80 kg/m^2) [46] and Portuguese players (189.9 ± 2.2 cm; 87.4 ± 8.7 kg; 24.5 kg/m^2) [47]. However, Asian players showed similar height (186.5 ± 4.4 cm) but lower body mass (80.8 ± 7.0 kg) and therefore lower BMI (23.3 kg/m^2) [48]. The same similarities can be seen between beach and team handball back players with Asian players being smaller and lighter [48]. The main difference between modalities lies in front players by which beach handball front players (wings and specialists) are as tall or taller than team handball players: 185.01 ± 5.46 cm, 184.2 ± 5.5 cm; 177.3 ± 5.0 cm but heavier and therefore with higher BMI: 84.66 ± 6.44 kg, 24.73 ± 1.49 kg/m^2; 81.6 ± 7.4 kg, 25.2 kg/m^2; 80.5 ± 6.1 kg, 25.7 kg/m^2; [46–48], respectively. An exception can be found in wings from the first league sample showing similar height, body mass and BMI values (176 ± 4 cm, 73.9 ± 4.2 kg, 23.9 ± 1.9 kg/m^2) than beach handball players of our study sample. The latter can be explained by the different demands for wings in beach handball compared to team handball. Due to the reduced playing area and the number of players (four in beach handball vs. seven in team handball), beach handball wings do not usually play as laterally as team handball wings.

Beach handball female players showed minor differences in height and body mass with respect to their Spanish team elite counterparts (171.31 ± 7.42 cm and 67.55 ± 8.06 kg) [23]. This trend can also be observed in studies comparing height and body mass in elite and sub-elite female players, with no statistically significant differences [49,50]. However, height and body mass differences were higher for Danish (175.1 ± 2.8 cm and 69.0 ± 6.2 kg) [51] or Norwegian female team handball players (179.0 ± 0.4 cm and 72.0 ± 6.3 kg) [52]. Other team sports, such as football, have also reported that Scandinavian female players are taller and heavier than Spanish female players [53,54]. Future studies should analyze the anthropometric profile of beach handball female players of other nationalities to confirm this tendency. In accordance with this, beach handball players showed lower BMI values than team handball players. Both with our study sample: 22.0 ± 1.5 kg/m^2 and with other elite beach handball female players from Spain: 21.68 ± 1.45 kg/m^2 [25] and from Brazil: 22.74 ± 2.5 kg/m^2 [10], BMI values were lower than female team handball players from Greece (23.6 ± 2.7 kg/m^2) [19], Czech Republic (23.4 ± 2.3 kg/m^2) [8]. This finding suggests that the lower values of basic anthropometric characteristics in beach handball players may play a role in the adaptation to the inherent features of the game, such as the resistance to displacement in sandy surfaces compared to court.

With regards to playing positions, our study sample showed similar values of height, body mass and BMI than Spanish team goalkeepers (174.96 ± 6.30 cm, 69.27 ± 7.66 kg, 22.60 ± 1.89 kg/m^2), wings (165.49 ± 4.83 cm, 61.23 ± 4.29 kg, 22.35 ± 1.13 kg/m^2) [23] and Danish back players (170.6 ± 5.0 cm, 65.2 ± 2.7 kg, 22.6 kg/m^2) [51]. However, although female beach handball back players showed similar height values than the aforementioned Spanish and Danish sample of back players (174.19 ± 6.21 cm and 175.1 ± 5.3 cm, respectively), our study sample was lighter and therefore had a lower BMI than team handball back players: 71.13 ± 7.8 kg, 23.44 ± 2.32 kg/m^2 and 71.4 ± 6.1 kg, 23.3 kg/m^2.

Regarding detailed anthropometric measures, the sum of 6 skinfolds value of the female sample is in accordance with other beach handball studies (84.50 ± 20.85 mm) [25] and lower than elite team handball players (95.50 ± 23.49 mm) [23], (92.20 ± 22.48 mm) [22]. Likewise for male players, the sum is lower than their team counterparts (77.2 ± 27.5 mm) [55]. Individual measures of skinfolds also show the same tendency by which elite team handball players show a greater amount of subcutaneous fat than beach handball.

The somatotype of our study sample showed that male players could be defined as balanced mesomorphic (2.6-4.4-2.7)—similar to team handball male players (3.01-4.85-2.29) [56] and soccer and volleyball players (1.6-4.7-2.9 and 2.0-4.0-3.2) [14]. Position-related somatotypes revealed similar mesomorphy and ectomorphy components than top-level team handball players: 4.51-2.45, 4.81-2.51 and 4.61-2.66 for goalkeepers, wings and backs, respectively [57]. The main difference lies in the endomorphy by which beach handball players show larger values for wings and backs than team handball (1.4 and 1.67), but especially for goalkeepers, who showed large differences (1.21). The larger endomorph component in goalkeepers could be explained by their more static play compared to other positions and the requirement of a large body area to cover the goal.

Beach handball female players showed a mean mesomorph-endomorph somatotype (3.5-3.3-2.6) with similar results to a study of female team handball players (3.06-2.53-2.64) [58] but fairly different to that of Greek players (4.2-4.7-1.8) [19]. Similarly to male beach handball players, position-related differences showed comparable ectomorphy components than female team handball players for the three positions, goalkeepers, wings and backs (2.72, 3.00 and 3.45) [58]. The difference with our female sample lies in the remaining components, by which endomorphy is similar to team handball: 3.50, 2.68 and 2.68 for goalkeepers, wings and backs, respectively, but mesomorphy is higher in beach handball: 3.0, 3.4 and 2.9 for goalkeepers, wings and backs, respectively. Larger levels of mesomorphy could be related to higher performance in short efforts and accelerations, which would have a positive transfer in beach handball performance [59].

In our study, male players showed body composition values lower than team handball players. Muscular mass was 36.2 ± 3.1 kg in the study sample whereas team counterparts reported 42.1 ± 7.9 kg [48] and 46.58 ± 4.25 kg [56]. Similarly, bone mass is lower in our study (13.3 ± 1.8 kg) than in the latter study on team handball. (18.02 ± 1.07 kg) [56]. Higher values of body composition components of team handball could be explained by the also higher values of weight and height discussed above. This tendency can also be observed for our female study sample with muscular mass (22.9 ± 1.5 kg) or another study with female beach handball players (22.44 ± 1.30 kg) [25] in comparison to female team handball players (25.01 ± 2.60 kg) [23]. Likewise, body fat mass percentage in our female study sample (15.4 ± 3.7%) is similar to other studies with beach handball (14.48 ± 3.06%) [25] and lower than female team handball players from Czech Republic (21.43 ± 2.48%) [8] and Greece (25.9 ± 3.3%) [19].

Our results showed that morphological characteristics should be taken into account to select players for individual positions with distinctive differences in male and female players. Professional coaches and researchers working within this specific sport should program their training strategies considering the general and position-specific tasks throughout the game. Similarly, the design of physical tests to specifically evaluate beach handball players in different playing positions could be based on the results obtained in this study, with special attention to those with large to extremely large effect sizes ($d > 1.2$).

Table 6. Summary table of studies examining age, height, body mass, BMI, body fat mass percentage and somatotype of beach and team handball players.

	Study	Level/Position	Discipline	Age (Years)	Height (cm)	Body Mass (kg)	BMI (kg/m^2)	Body Fat (%)	Somatotype
Male Players	Zapardiel et al. [5]	Spain Elite	Beach	25.3 ± 4.8	187.5 ± 7.5	87.0 ± 9.5	24.9 [a]	-	-
	Ghobadi et al. [46]	Spain Elite	Team	28.2 ± 4.0	192.9 ± 7.6	96.9 ± 11.2	26.0 ± 2.4	-	-
	Ghobadi et al. [46]	World Elite	Team	26.9 ± 4.2	190.1 ± 6.8	92.4 ± 9.8	25.5 ± 2.1	-	-
	Šibila and Pori [56]	Slovenian League	Team	25.1 ± 4.3	188.4 ± 5.5	89.6 ± 8.4	25.3 [a]	11.3 ± 2.4	3.0-4.8-2.3
	Present study	Spain Elite	Beach	27.1 ± 5.2	187.4 ± 8.2	85.2 ± 11.3	24.2 ± 2.5	11.7 ± 3.9	2.6-4.4-2.7
		Front	Beach	26.1 ± 5.5	181.4 ± 6.4	77.1 ± 7.2	23.5 ± 2.4	10.3 ± 2.3	2.1-5.0-2.7
		Back	Beach	28.2 ± 5.8	194.1 ± 5.7	93.4 ± 10.7	24.8 ± 2.9	12.8 ± 5.1	2.9-4.0-2.8
		Goalkeepers	Beach	27.5 ± 3.5	187.9 ± 2.0	88.6 ± 0.8	25.0 ± 0.8	13.2 ± 4.4	4.1-3.6-2.3
Female Players	Becerra et al. [25]	Spain Elite	Beach	22.9 ± 4.0	167.9 ± 4.4	61.0 ± 4.0	21.7 ± 1.4	14.5 ± 3.1	3.3-3.3-2.6
	Zapardiel et al. [5]	Spain Elite	Beach	25.3 ± 4.8	168.0 ± 3.9	60.8 ± 3.9	21.5 [a]	-	-
	Sena et al. [10]	Brazil Elite	Beach	26.8 ± 7.8	169.5 ± 8.1	65.4 ± 9.4	22.7 ± 2.5	22.0 ± 3.2	4.0-4.6-2.3
	Silva et al. [26]	Brazil World Champ	Beach	24.7 ± 2.0	168.0 ± 10.0	63.8 ± 7.1	-	-	-
	Vila et al. [23]	Spain League	Team	25.7 ± 4.5	171.3 ± 7.4	67.5 ± 8.1	23.0 ± 1.7	-	3.9-4.3-2.3
	Michalsik et al. [51]	Denmark League	Team	25.3 ± 6.0	175.1 ± 2.8	69.0 ± 6.2	22.5 [a]	-	-
	Ronglan et. al. [52]	Norway National	Team	23.7 ± 2.1	179.0 ± 0.4	72.0 ± 6.3	22.5 [a]	-	-
	Bayios et al. [19]	Greece League	Team	21.5 ± 4.6	165.9 ± 6.3	65.1 ± 9.1	23.6 ± 2.7	25.9 ± 3.3	4.2-4.7-1.8
	Mala et al. [8]	Check Rep. National	Team	24.0 ± 3.5	176.0 ± 6.5	72.5 ± 8.3	23.4 ± 2.3	21.4 ± 2.5	-
	Cavala and Katic [58]	Croatian League	Team	-	-	-	-	-	3.1-2.5-2.6
	Present study	Spain World Champ	Beach	24.1 ± 4.7	169.1 ± 5.1	62.9 ± 5.3	22.0 ± 1.5	15.4 ± 3.7	3.5-3.3-2.6
		Front	Beach	22.9 ± 4.5	167.1 ± 4.1	61.5 ± 4.0	22.0 ± 1.3	15.0 ± 3.0	3.5-3.4-2.4
		Back	Beach	23.0 ± 1.4	171.7 ± 5.4	58.5 ± 3.9	19.8 ± 0.1	12.0 ± 2.7	2.6-2.9-3.8
		Goalkeepers	Beach	28.7 ± 5.1	174.0 ± 5.1	70.3 ± 2.1	23.3 ± 1.4	19.1 ± 4.3	4.1-3.0-2.3

BMI: Body Mass Index, Somatotype (endomorphy-mesomorphy-ectomorphy), [a] Data computed by authors from mean height and body mass values, - = Data not available.

4.1. Limitations

The main limitation of this study is the small sample size, especially regarding positional playing subgroups. Therefore, the findings of the study should be interpreted with caution. Although being world-class elite beach handball players with very little prior research, the sample size is not powerful enough to support statements from anthropometrics and positional success for the entire population of beach handball players. Furthermore, a possible bias derived from an exclusively Spanish sample must be taken into account. The only studies with female world champion samples indicate small variations in most body measures, so the beach handball elite profile could start to converge to stable values, regardless of nationality. Still, future studies are required to confirm our results in a larger sample, particularly including other playing levels and nationalities to investigate definitive player profiles.

4.2. Practical Applications

The practical application of our study can be found in results with large ($>$1.2) to extremely large ($>$4.0) effect sizes. The most remarkable results from a practical point of view are basic anthropometric characteristics between genders (body height, body mass, arm span, and marginally BMI ($d \approx 1.2$), which can be analyzed against other related studies in Table 6. Arm skinfolds, girths and breadths, and muscular mass are distinctive differences between male and female players. The position-related analysis revealed similar practical meaningful differences for male players (body height, body mass, and arm span) and for female players (body mass, BMI, and muscular mass) between playing positions. Results with lower d values would have needed a larger sample to provide reliable evidence. However, with the above limitations, these findings provide reference values for beach handball players that could help coaches to accurately control training to improve athletes' performance and to identify young elite players. Additionally, the data from this study can be used to design larger confirmatory studies that would expand the findings of this research. It is therefore recommended to implement anthropometric measures and somatotype determination to confirm which anthropometric factors that would contribute to performance at specific playing positions.

5. Conclusions

This study has examined anthropometric characteristics, somatotype and body composition of elite beach handball players and compare male and female players to highlight differences between groups. Anthropometric characteristics, girths, breadths, and absolute components of body composition were higher in male players than female players. Conversely, female players showed larger values of skinfolds and body fat mass percentage. The mean somatotype for male and female players was balanced mesomorph and endomorphic-mesomorph, respectively. According to position, the optimal somatotype rating for elite beach handball players would range as mesomorph-endomorph for male and mesomorphic endomorph for female goalkeepers, balanced mesomorph for male and balanced ectomorph for female back players and ectomorphic mesomorph for male and mesomorph-endomorph for female front players. These findings provide reference values for beach handball players that could help coaches to accurately control training to improve athletes' performance and to identify young elite players. However, due to the low sample size, especially for position-related measures, further profile studies of beach handball players are required before establishing a definitive reference framework.

Author Contributions: Conceptualization, B.P., J.J.E.-A. and A.P.-T.; Data curation, J.J.E.-A. and S.S.-P.; Formal analysis, B.P., S.S.-P. and A.P.-T.; Investigation, B.P. and S.S.-P.; Methodology, J.J.E.-A., S.S.-P. and A.P.-T.; Project administration, B.P.; Resources, J.J.E.-A. and A.P.-T.; Software, B.P. and A.P.-T.; Supervision, B.P.; Validation, B.P. and A.P.-T.; Visualization, S.S.-P.; Writing—original draft, B.P.; Writing—review and editing, B.P., J.J.E.-A., S.S.-P. and A.P.-T. All authors have read and agreed to the published version of the manuscript.

Funding: This research received no external funding.

Conflicts of Interest: The authors declare no conflict of interest.

References

1. Gorostiaga, E.M.; Granados, C.; Ibáñez, J.; Izquierdo, M. Differences in physical fitness and throwing velocity among elite and amateur male handball players. *Int. J. Sports Med.* **2005**, *26*, 225–232. [CrossRef] [PubMed]
2. Karcher, C.; Buchheit, M. On-Court demands of elite handball with special reference to playing positions. *Sports Med.* **2014**, *44*, 797–814. [CrossRef] [PubMed]
3. Michalsik, L.B.; Aagaard, P. Physical demands in elite team handball: Comparisons between male and female players. *J. Sports Med. Phys. Fit.* **2015**, *55*, 878–891.
4. Pueo, B.; Jimenez-Olmedo, J.M.; Penichet-Tomas, A.; Ortega Becerra, M.; Espina Agullo, J.J. Analysis of time-motion and heart rate in elite male and female beach handball. *J. Sports Sci. Med.* **2017**, *16*, 450–458.
5. Zapardiel, J.C.; Asín-Izquierdo, I. Conditional analysis of elite beach handball according to specific playing position through assessment with GPS. *Int. J. Perform. Anal. Sport* **2020**, *20*, 118–132. [CrossRef]
6. Chelly, M.S.; Hermassi, S.; Shephard, R.J. Relationships between power and strength of the upper and lower limb muscles and throwing velocity in male handball players. *J. Strength Cond. Res.* **2010**, *24*, 1480–1487. [CrossRef]
7. Bon, M.; Pori, P.; Sibila, M. Position-related differences in selected morphological body characteristics of Top-level female handball players. *Coll. Antropol.* **2015**, *39*, 631–639.
8. Mala, L.; Maly, T.; Zahalka, F.; Bunc, V.; Kaplan, A.; Jebavy, R.; Tuma, M. Body composition of elite female players in five different sports games. *J. Hum. Kinet.* **2015**, *45*, 207–215. [CrossRef]
9. Massuça, L.; Fragoso, I. Morphological characteristics of adult male handball players considering five levels of performance and playing position. *Coll. Antropol.* **2015**, *39*, 109–118.
10. Sena, J.E.A.; Gomes, A.L.M.; Mimbacas, A.; Ferreira, U.M.G. Dermatoglifia, somatotipo e composição corporal no beach handball: Estudo comparativo entre diferentes níveis de qualificação esportiva. *Motricidade* **2012**, *8*, 567–576.
11. Manchado, C.; Tortosa-Martínez, J.; Vila, H.; Ferragut, C.; Platen, P. Performance factors in women's team dhandball: Physical an physiological aspects–A review. *J. Strength Cond. Res.* **2013**, *27*, 1708–1719. [CrossRef] [PubMed]
12. Van Den Tillaar, R.; Cabri, J.M.H. Gender differences in the kinematics and ball velocity of overarm throwing in elite team handball players. *J. Sports Sci.* **2012**, *30*, 807–813. [CrossRef]
13. Zapartidis, I.; Skoufas, D.; Vareltzis, I.; Christodoulidis, T.; Toganidis, T.; Kororos, P. Factors influencing ball throwing velocity in young female handball players. *Open Sports Med. J.* **2009**, *3*, 39–43. [CrossRef]
14. Campa, F.; Silva, A.M.; Talluri, J.; Matias, C.N.; Badicu, G.; Toselli, S. Somatotype and bioimpedance vector analysis: A new target zone for male athletes. *Sustainability* **2020**, *12*, 4365. [CrossRef]
15. Granados, C.; Izquierdo, M.; Ibañez, J.; Bonnabau, H.; Gorostiaga, E.M. Differences in physical fitness and throwing velocity among elite and amateur female handball players. *Int. J. Sports Med.* **2007**, *28*, 860–867. [CrossRef]
16. Van den Tillaar, R.; Ettema, G. Effect of body size and gender in overarm throwing performance. *Eur. J. Appl. Physiol.* **2004**, *91*, 413–418. [CrossRef]
17. Debanne, T.; Laffaye, G. Predicting the throwing velocity of the ball in handball with anthropometric variables and isotonic tests. *J. Sports Sci.* **2011**, *29*, 705–713. [CrossRef]
18. Ackland, T.R. Built for Success: Homogeneity in Elite Athlete Morphology. In *Kinanthropometry IX: Proceedings of the 9th International Conference of the International Society for the Advancement of Kinanthropometry*; Marfell-Jones, M., Olds, T., Stewart, A., Eds.; Routledge: London, UK, 2006; pp. 26–34.
19. Bayios, I.A.; Bergeles, N.K.; Apostolidis, N.G.; Noutsos, K.S.; Koskolou, M.D. Anthropometric, body composition and somatotype differences of Greek elite female basketball, volleyball and handball players. *J. Sports Med. Phys. Fit.* **2006**, *46*, 271–280.
20. Hasan, A.A.; Reilly, T.; Cable, N.T.; Ramadan, J. Anthropometric profiles of elite Asian female handball players. *J. Sports Med. Phys. Fit.* **2007**, *47*, 197–202.
21. Saavedra, J.M.; Kristjánsdóttir, H.; Einarsson, I.; Guðmundsdóttir, M.L.; Þorgeirsson, S.; Stefansson, A. Anthropometric characteristics, physical Fitness, and throwing velocity in elite women's handball teams. *J. Strength Cond. Res.* **2018**, *32*, 2294–2301. [CrossRef]
22. Moss, S.L.; McWhannell, N.; Michalsik, L.B.; Twist, C. Anthropometric and physical performance characteristics of top-elite, elite and non-elite youth female team handball players. *J. Sports Sci.* **2015**, *33*, 1780–1789. [CrossRef] [PubMed]
23. Vila, H.; Manchado, C.; Rodriguez, N.; Abraldes, J.A.; Alcaraz, P.E.; Ferragut, C. Anthropometric profile, vertical jump, and throwing velocity in elite female handball players by playing positions. *J. Strength Cond. Res.* **2012**, *26*, 2146–2155. [CrossRef]

24. Ziv, G.; Lidor, R. Physical attributes, physiological characteristics, on-court performances and nutritional strategies of female and male basketball players. *Sports Med.* **2009**, *39*, 547–568. [CrossRef] [PubMed]
25. Becerra, M.O.; Espina-Agulló, J.J.; Pueo, B.; Jiménez-Olmedo, J.M.; Penichet-Tomás, A.; Sellés-Pérez, S. Anthropometric profile and performance indicators in female elite beach handball players. *J. Phys. Educ. Sport* **2018**, *18*, 1155–1160.
26. Silva, A.S.; Coeli Seabra Marques, R.; DE Azevedo Lago, S.; Guedes Santos, D.A.; Lacerda, L.M.; Silva, D.C.; Soares, Y.M. Physiological and nutritional profile of elite female beach handball players from Brazil. *J. Sports Med. Phys. Fit.* **2016**, *56*, 503–509.
27. Knechtle, B.; Wirth, A.; Baumann, B.; Knechtle, P.; Rosemann, T.; Oliver, S. Differential correlations between anthropometry, training volume, and performance in male and female Ironman triathletes. *J. Strength Cond. Res.* **2010**, *24*, 2785–2793. [CrossRef]
28. Nikolaidis, P.T.; Rosemann, T.; Knechtle, B. Skinfold thickness distribution in recreational marathon runners. *Int. J. Environ. Res. Public Health* **2020**, *17*, 2978. [CrossRef]
29. Bøgild, P.; Jensen, K.; Kvorning, T. Physiological performance characteristics of danish national team handball players 1990–2016: Implications on position-specific strength and conditioning training. *J. Strength Cond. Res.* **2019**, *34*, 1555–1563. [CrossRef]
30. Hermassi, S.; Laudner, K.; Schwesig, R. Playing level and position differences in body characteristics and physical fitness performance among male team handball players. *Front. Bioeng. Biotechnol.* **2019**, *7*, 149. [CrossRef]
31. Schwesig, R.; Hermassi, S.; Fieseler, G.; Irlenbusch, L.; Noack, F.; Delank, K.S.; Shephard, R.J.; Chelly, M.S. Anthropometric and physical performance characteristics of professional handball players: Influence of playing position. *J. Sports Med. Phys. Fit.* **2017**, *57*, 1471–1478.
32. Ross, W.D.; Marfell-Jones, M.J. Kinanthropometry. In *Physiological Testing of Elite Athlete*; Human Kinetics Publishers Inc.: London, UK, 1991; pp. 223–308.
33. Withers, R.T.; Craig, N.P.; Bourdon, P.C.; Norton, K.I. Relative body fat and anthropometric prediction of body density of male athletes. *Eur. J. Appl. Physiol. Occup. Physiol.* **1987**, *56*, 191–200. [CrossRef] [PubMed]
34. Lee, R.C.; Wang, Z.; Heo, M.; Ross, R.; Janssen, I.; Heymsfield, S.B. Total-body skeletal muscle mass: Development and cross-validation of anthropometric prediction models. *Am. J. Clin. Nutr.* **2000**, *72*, 796–803. [CrossRef] [PubMed]
35. Rocha, M. Peso ósseo do brasileiro de ambos os sexos de 17 a 25 anhos. *Arq. Anat. Antropol.* **1975**, *1*, 445–451.
36. Cruz, J.R.A.; Armesilla, M.D.C.; De Lucas, A.H.; Riaza, L.M.; Pascual, C.M.; Manzañido, J.P.; Quintana, M.S.; Belando, J.E.S. Protocolo de valoración de la composición corporal para el reconocimiento médico-deportivo. documento de consenso del grupo español de cineantropometría (GRXC) de la federación española de medicina del deporte (FEMEDE). Versión 2010. *Arch. Med. Deport.* **2010**, *27*, 330–344.
37. Heath, B.H.; Carter, J.E. A modified somatotype method. *Am. J. Phys. Anthropol.* **1967**, *27*, 57–74. [CrossRef]
38. Carter, J.E.L.; Heath, B.H. *Somatotyping: Development and Applications*; Cambridge University Press: Cambridge, NY, USA, 1990.
39. Cohen, J. *Statistical Power Analysis for the Behavioral Sciences*; Routledge: England, UK, 1988.
40. Hopkins, W.G. A Scale of Magnitudes for Effect Statistics. A New View of Statistics. Available online: www.sportsci.org/resource/stats/effectmag.html (accessed on 1 March 2020).
41. Garcia-Gil, M.; Torres-Unda, J.; Esain, I.; Duñabeitia, I.; Gil, S.M.; Gil, J.; Irazusta, J. Anthropometric parameters, age, and agility as performance predictors in elite female basketball players. *J. Strength Cond. Res.* **2018**, *32*, 1723–1730. [CrossRef]
42. Popovic, S.; Akpinar, S.; Jaksic, D.; Matic, R.; Bjelica, D. Comparative study of anthropometric measurement and body composition between elite soccer and basketball players. *Int. J. Morphol.* **2013**, *32*, 461–467. [CrossRef]
43. Popovic, S.; Bjelica, D.; Jaksic, D.; Hadzic, R. Comparative study of anthropometric measurement and body composition between elite soccer and volleyball players. *Int. J. Morphol.* **2014**, *32*, 267–274. [CrossRef]
44. Pietraszewska, J.; Burdukiewicz, A.; Stachoń, A.; Andrzejewska, J.; Pietraszewski, B. Anthropometric characteristics and lower limb power of professional female volleyball players. *S. Afr. J. Res. Sport Phys. Educ. Recreat.* **2015**, *37*, 99–111.
45. Ingebrigtsen, J.; Dillern, T.; Shalfawi, S.A. Aerobic capacities and anthropometric characteristics of elite female soccer players. *J. Strength Cond. Res.* **2011**, *25*, 3352–3357. [CrossRef]

46. Ghobadi, H.; Rajabi, H.; Farzad, B.; Bayati, M.; Jeffreys, I. Anthropometry of world-class elite handball players according to the playing position: Reports from men's handball world championship. *J. Hum. Kinet.* **2013**, *39*, 213–220. [CrossRef] [PubMed]

47. Póvoas, S.C.A.; Ascensaõ, A.A.M.R.; Magalhães, J.; Seabra, A.F.; Krustrup, P.; Soares, J.M.C.; Rebelo, A.N.C. Physiological demands of elite team handball with special reference to playing position. *J. Strength Cond. Res.* **2014**, *28*, 430–442. [CrossRef] [PubMed]

48. Hasan, A.A.A.; Rahaman, J.A.; Cable, N.T.; Reilly, T. Anthropometric profile of elite male handball players in Asia. *Biol. Sport* **2007**, *24*, 3–12.

49. Milanese, C.; Piscitelli, F.; Lampis, C.; Zancanaro, C. Anthropometry and body composition of female handball players according to competitive level or the playing position. *J. Sports Sci.* **2011**, *29*, 1301–1309. [CrossRef] [PubMed]

50. Milanese, C.; Piscitelli, F.; Lampis, C.; Zancanaro, C. Effect of a competitive season on anthropometry and three-compartment body composition in female handball players. *Biol. Sport* **2012**, *29*, 199–204. [CrossRef]

51. Michalsik, L.B.; Aagaard, P.; Madsen, K. Technical activity profile and influence of body anthropometry on playing performance in female elite team handball. *J. Strength Cond. Res.* **2015**, *29*, 1126–1138. [CrossRef]

52. Ronglan, L.T.; Raastad, T.; Børgesen, A. Neuromuscular fatigue and recovery in elite female handball players. *Scand. J. Med. Sci. Sport* **2006**, *16*, 267–273. [CrossRef]

53. Andersson, H.Å.; Randers, M.B.; Heiner-Møller, A.; Krustrup, P.; Mohr, M. Elite female soccer players perform more high-intensity running when playing in international games compared with domestic league games. *J. Strength Cond. Res.* **2010**, *24*, 912–919. [CrossRef]

54. Sedano Campo, S.; Cuadrado Sáenz, G.; Redondo Castán, J.; Benito Trigueros, A. Anthropometric profile of spanish female soccer players analysis depending on the competition level and the usual playing position in the field. *Apunt. Educ. Física Deport.* **2009**, *2009*, 78–87.

55. Garrido-Chamorro, R.; Sirvent-Belando, J.E.; González-Lorenzo, M.; Blasco-Lafarga, C.; Roche, E. Skinfold sum: Reference values for top athletes. *Int. J. Morphol.* **2012**, *30*, 803–809. [CrossRef]

56. Šibila, M.; Pori, P. Position-related differences in selected morphological body characteristics of top-level handball players. *Coll. Antropol.* **2009**, *33*, 1079–1086. [PubMed]

57. Urban, F.; Kandráč, R.; Táborský, F. Position-Related Categorization of Somatotypes in Top Level Handball Players. 2014. Available online: https://www.researchgate.net/publication/222108359_Position-Related_Categorization_Of_Somatotypes_In_Top_Level_Handball_Players (accessed on 18 August 2020).

58. Cavala, M.; Katic, R. Morphological, Motor and Situation-Motor Characteristics of Elite Female Handball Players According to Playing Performance and Position. *Coll. Antropol.* **2010**, *34*, 1355–1361. [PubMed]

59. Aerenhouts, D.; Delecluse, C.; Hagman, F.; Taeymans, J.; Debaere, S.; Van Gheluwe, B.; Clarys, P. Comparison of anthropometric characteristics and sprint start performance between elite adolescent and adult sprint athletes. *Eur. J. Sport Sci.* **2012**, *12*, 9–15. [CrossRef]

 *sustainability*

Article

The Effectiveness of Biological Maturation and Lean Mass in Relation to Muscle Strength Performance in Elite Young Athletes

Paulo Francisco de Almeida-Neto [1],*, Dihogo Gama de Matos [2], Adam D. G. Baxter-Jones [3], Gilmário Ricarte Batista [4], Vanessa Carla Monteiro Pinto [1], Matheus Dantas [1], Felipe J. Aidar [2,5], Paulo Moreira Silva Dantas [1] and Breno Guilherme de Araújo Tinoco Cabral [1]

[1] Department of Physical Education, Federal University of Rio Grande do Norte, Natal 59078-970, Brazil; vanessa.pinto@unp.br (V.C.M.P.); matheusp_dantas@ufrn.edu.br (M.D.); pgdantas@ufrnet.br (P.M.S.D.); brenotcabral@reitoria.ufrn.br (B.G.d.A.T.C.)
[2] Group of Studies and Research of Performance, Sport, Healt and Paralympic Sports GEPEPS, the Federal University of Sergipe, UFS, São Cristovão 49100-000, Sergipe, Brazil; dihogo.dematos@mail.mcgill.ca (D.G.d.M.); fjaidar@academico.ufs.br (F.J.A.)
[3] College of Kinesiology, University of Saskatchewan, Saskatoon, SK S7N 5B2, Canada; baxter.jones@usask.ca
[4] Department of Physical Education, Federal University of Paraíba, João Pessoa 58051-900, Brazil; cajagr@ccs.ufpb.br
[5] Department of Physical Education, Federal University of Sergipe, São Cristovão 49100-000, Sergipe, Brazil
* Correspondence: paulojitte@ufrn.edu.br or paulo220911@hotmail.com; Tel.: +55-849-8188-8269

Received: 17 July 2020; Accepted: 8 August 2020; Published: 19 August 2020

Abstract: This study aimed to identify the interactional relationships between maturation (biological age (BA)) and lean mass on strength development in young athletes from different sports. Using a cross-sectional study design, a sample of 64 young athletes (rowers, swimmers, jiu-jitsu, volleyball, soccer and tennis players) of both sexes (13.6 ± 1.17 years) were recruited. Body composition was assessed using dual energy bone densitometry with X-ray source (DEXA). Strength of upper limbs (ULS), force hand grip (HG), vertical jump (VJ) and jump against movement (CMJ) were recorded. BA was estimated from anthropometrics. BA relationships were identified with upper limb strength in all athletes, and with the lower limb strength of tennis players, only ($p < 0.05$). An interaction effect between lean mass and BA was found ($\eta^2 p = 0.753$), as was a local effect within the regression models ($f^2 \geq 0.33$). Athletes with a higher concentration of lean mass had superior upper and lower limb strength ($p < 0.05$). Lean mass showed a local effect (f^2) greater than that associated with BA. Although maturation is related to strength development, the strength of the relationship is mitigated by the accrual of lean mass. Specifically, the local effect of lean mass on muscle strength is broader than that of maturation, especially for lower limb strength.

Keywords: young athletes; lean mass; maturation; performance sensory motor; sports; puberty

---

1. Introduction

Numerous factors determine success in high-performance sport, including psychological conditions, motor skills and body composition [1]. Strength has been linked to performance levels and, in childhood, it has been used to identify talent in a variety of sports [2–6]. Previous reports have shown that selected young athletes, within chronological age (CA) bands, are stronger than those deselected [4,5]. In addition, those selected have been observed to be more advanced in their maturation (indexed by biological age (BA), i.e., more mature then their CA matched peers) [5].

There is great variation in the timing and tempo of maturation [5,6]. Within individual's maturation stages can be classified as late, synchronized (average) or accelerated (early) in relation to chronological age [6]. Researchers have observed interaction effects between CA and BA in a variety of performance tasks such as: acceleration time, sprint speed, speed with change of direction and jumping height [7]. In addition, Charcharis et al. [8] demonstrated that when strength was normalized to body weight, there were still differences between average and early maturers and late maturers in strength outcomes.

The voluntary production of movements in the neuromuscular system activates the nervous and locomotor systems to perform tasks [9]. Movement occurs due to the interaction of muscle tissue with bone tissue, with measures of lean mass being the determining factor for the execution of body movement [10]. Increased volume of lean mass is also identified as a significant characteristic for physical health and sports performance [11,12]. The amount of lean mass has been shown to be a strong indicator of performance in young athletes including, but not limited to: muscle strength and aerobic and anaerobic capacity [4,13–15].

Individuals with a greater amount of lean mass have a greater number of sarcomeres (i.e., biological units where muscle tension occurs) which promotes the production of strength through muscle contraction [4,14,16]. In relation to sport, the particularities of each modality generally require a specific manifestation of muscle strength (endurance strength or muscle power), different from the unparticular movements normally used [8,17,18]. For example, Azimi et al. [4] identified that the muscular strength of young elite Iranian weight lifters was influenced by the accumulation of lean mass, however, the authors did not analyze the possible interaction of biological maturation, which could have confounded this interpretation. In addition, other studies have identified that BA, CA and sex all have independent significant relationships with the production of muscle strength in young athletes and non-athletes [19–23].

Thus, the type of stimuli received by the characteristics of sports can significantly influence the levels of strength in the upper and lower limbs. For example, volleyball and basketball athletes constantly perform jumps, which can favor increased strength in the lower limbs. Rowers, tennis players and javelin throwers perform movements with an emphasis on upper limb usage which can enhance the strength in these upper limbs [1–4,8,16,21].

Although the independent effects of biological maturation and lean mass on muscle strength are well known, there is scarce information about the interactive relationship of BA and lean mass accrual on muscular strength development in young athletes [6,24–26]. Thus, the objective of the present study was to study the independent and interactive relationships between biological age and lean mass on muscular strength development of the upper and lower limbs in young athletes.

We hypothesized that: (i) Muscular strength of the upper and lower limbs is different between sports when the confounders of sex, sport and size are controlled; (ii) maturation (BA) would significantly interact with levels of lean mass; (iii) young athletes with higher concentrations of lean mass, would have an older BA and better performance of upper and lower limb muscle strength; and (iv) lean mass would have a greater local effect on the muscular strength of upper and lower limbs in relation to biological maturation.

2. Methods

Participants

The sample consisted of 64 young athletes, between 12 and 16 years of age, from both sexes (73% male and 27% female), recruited from six different sports: rowing (17%, 9 male and 2 female), swimming (17%, 7 male and 4 female), volleyball (16%, 2 male and 8 female), soccer (17%, 11 male), tennis (17%, 9 male and 2 female) and jiu-jitsu (16%, 9 male and 1 female). The average age of the sample was 13.6 ± 1.17 years. According to the state federation of each sport, all athletes were ranked among the top 15 for their age group and had been nominated for the best athletes of the year for 2019 by the Brazilian confederation of their respective sports. Based on information from national confederations, the sample can be classified as national level athletes (high level sports), in accordance with Matsudo et al. [3].

The sample size was determined a priori, so we adopted an $\alpha < 0.05$ and a $\beta = 0.80$, estimating an effective size = 0.80. Subsequently, we reached the number of 64 subjects for the sample. There was no sample loss. The participants were members of sports clubs in the city of Natal, Brazil. The inclusion criteria were: (1) aged between 12 and 16 years, (2) be registered with a federation of their respective sport and (3) have a weekly training frequency of more than three days. The exclusion criteria were: (1) use of exogenous substances that could influence the performance of muscle strength, (2) health problem that could interfere with the results and (3) not agreeing to participate in all stages of the research.

The research was approved by the Ethics and Research Committee of the Federal University of Rio Grande do Norte—Brazil (CAE: 15865619.7.0000.5537; Opinion: 3.552.010) according to Resolution 466/12 of the National Health Council, on 12 December 2012, strictly respecting the national and international ethical principles contained in the Declaration of Helsinki. It is worth mentioning that the study complied with all the international requirements and standards of the STROBE checklist for observational studies of incidence and prevalence [27].

3. Anthropometry

Anthropometric assessments were performed according to the International Society of the Advancement of Kinanthropometry protocols [28]. Body mass was measured using a digital scale with an accuracy of 0.1 kg (FILIZOLA®, São Paulo, Brazil). Height and sitting height (trunk length) were assessed using a stadiometer with an accuracy of 0.01 cm (SANNY®, São Paulo, Brazil). Leg length was calculated by subtracting trunk length from height [29]. The perimeter was measured using an anthropometric tape (SANNY®, São Paulo, Brazil), and the bone diameters were measured using a caliper (SANNY®, São Paulo, Brazil).

4. Maturation Analysis

Somatic maturation (years from attainment of peak height velocity (PHV); termed maturity offset) an assessment of maturity was estimated from anthropometric measures [29], using the following equations.

$$\text{Maturity offset in males} = -9.236 + [0.0002708 \times (\text{Leg length} \times \text{Trunk Height})] + [-0.001663 \times (\text{Age} \times \text{Leg length})] + [0.007216 \times (\text{Age} \times \text{Trunk Height})] + [0.02292 \times (\text{Weight/height}) \times 100] \tag{1}$$

$$\text{Maturity offset in females} = -9.376 + [0.0001882 \times (\text{Leg length} \times \text{Trunk Height})] + [0.0022 \times (\text{Age} \times \text{Leg length})] + [0.005841 \times (\text{Age} \times \text{Trunk Height})] - [0.002658 \times (\text{Age} \times \text{Weight})] + [0.07693 \times (\text{Weight/height}) \times 100] \tag{2}$$

Age at PHV was calculated as age at measurement—maturity offset. Three maturity categories were identified: (1) Pre-PHV (Maturity offset < -1); (2) circum-PHV (Maturity offset ≥ -1 and Maturity offset $\leq +1$); (3) post-PHV (Maturity offset $> +1$).

5. Body Composition

Body composition was assessed by a member of the research team with prior training through the use of dual energy bone densitometry with X-ray source (DEXA) (LUNAR®/GE PRODIGY—LNR 41,990, United States). This procedure is considered one of the most reliable standards for measuring body composition [30]. Through the use of appropriate algorithms for the pediatric population, DEXA analysis provided data on total fat and lean body mass (i.e., bone mineral density, bone mineral content and total lean mass).

6. Analysis of Muscular Performance of the Upper Limbs

The upper limbs strength (ULS) was assessed using the medicine ball test [31]. The subject was seated with his back against a wall and his knees extended. At the evaluator's signal, a medicine ball (Ax Sports®, Tangará, Brazil) with a mass of 2 kg positioned at the height of the sternum was thrown horizontally with both hands. Trunk movement was not allowed. The test was performed three consecutive times interspersed with a passive recovery period of three minutes. The best attempt was retained for analysis. The test was performed following the protocol described by Mello et al. [31], where all the subjects launched a medicine ball of 2 kg in weight, normalizing the weight of the medicine ball (i.e., ideal tension) is not considered. The test follows the logic that if two subjects launch an object of the same mass, the one that launches the most distant object has the advantage of ULS [31]. The distance reached at launch was attributed to the evaluated ULS capacity. The handgrip (HG) test was used to assess the strength of the upper limbs using a hydraulic dynamometer (JAMAR®, Cambuci, Brazil; calibrated before each assessment). The subjects remained seated on a bench with adjustable height and the forearm flexed at an angle of 90° [32]. All participants performed three maximum voluntary contractions (3 s) with the left and right limbs, interspersed with recovery periods of 60 s, and the best performance was retained for statistical analysis [32].

7. Analysis of Muscle Performance of the Lower Limbs

The performance of the lower limbs was attained by tests of vertical jump (VJ) and jump against movement (CMJ), both jumps were analyzed using a force platform (CEFISE®, São Paulo, Brazil). The protocols established by Forza and Edmundson [33] were utilized. A familiarization session was held with the techniques of the movements of VJ and CMJ, seeking to reduce errors during the execution of the protocols. Then, starting from an orthostatic position, held for three seconds, with the knees flexed at approximately 90° and the hands fixed on the waist, the subjects were instructed to perform a vertical jump as high as possible. For CMJ analysis the same recommendations were adopted, however, the volunteers performed a squat followed by the jump. A 10-minute recovery interval was established between VJ and CMJ. For both tests, three attempts were made, interspersed with 60 s of passive recovery and the best attempt was retained for data analysis.

8. Statistics

The normality of the data was tested for using the Kolmogorov-Smirnov test and z-score of asymmetry and kurtosis (−1.96 to 1.96). For the correlations we used the Pearson test. For partial correlations, we controlled the power of the lean mass variable, with the following magnitudes identified: insignificant: $r < 0.10$; weak: $r = 0.10$–0.39; moderate: $r = 0.40$–0.69; strong: $r = 0.70$–0.89; very strong: $r = 0.90$–1.00 [34]. To group variables with similar patterns, we used an analysis based on the unsupervised machine learning technique with k-means cluster through a Hartigan-Wong algorithm [35]. The groups formed by these analyses were used in the regression models, the confounding factors sex (male and female) and chronological age (CA) present in the sample were controlled during the regressions, and backdoor arithmetic was used to inhibit the effect of confounding factors during statistical analysis [36,37]. The homogeneity of the regression models was tested by the Breush-Pegan test and the assumptions of normality, variance and independence of the data were not denied. The size of the effect in the regressions was measured by the "r^2" coefficient. The explanatory power of the dependent variable was based on the independent variables. To test the multicollinearity of the regression models, the Durbin Watson test was used. Bonferroni's correction was applied, and later comparisons were performed. An ANCOVA test was used to control the confounding factor of sex and chronological age.

Bonferroni's post hoc test was used to find differences. The percentage of the variation coefficient (CV%) was calculated for the maturation variable using the formula: $CV\% = $ (standard deviation (SD)/mean) $\times 100$. The partial Eta-square ($\eta^2 p$) was used to verify the size of the effect of the interaction between the variances of the variables: [maturation × lean mass]; [maturation × height]; [maturation × body weight]; [lean mass × HG]; [lean mass × ULS]; [lean mass × VJ]; [lean mass × CMJ]; [HG × maturation]; [maturation × ULS]; [maturation × VJ]; [maturation × CMJ]. The magnitude of $\eta^2 p$ ranges from 0 to 1 and was interpreted as follows: Small $\eta^2 p \leq 0.10$ to 0.23; Average $\eta^2 p$ from 0.24 to 0.34; Large $\eta^2 p$ from 0.35 to 0.44; Very large $\eta^2 p \geq 0.45$ [38]. Through Cohen's f^2, the local effect of the variables was verified in relation to: [lean mass × HG]; [lean mass × ULS]; [lean mass × VJ]; [lean mass × CMJ]; [HG × maturation]; [maturation × ULS]; [maturation × VJ]; [maturation × CMJ]; [maturation × lean mass]. The magnitude of f^2 ranges from 0 to 1 and was interpreted as follows: $f^2 \leq 0.02$ to 0.14; medium: $f^2 \geq 0.15$ to 0.34; large: $f^2 \geq 0.35$ [39]. All analyses were performed using open source software R (version 4.0.1; R Foundation for Statistical Computing®, Vienna, Austria) considering $p < 0.05$.

9. Results

Table 1 shows the descriptive characterization of the sample. The average age of the subjects was were 13.6 ± 1.17 years with an average age of PHV of 13.5 ± 1.82 years (95% between 9.9 and 17.1 years). The average maturity offset was 0.14 ± 1.82 years from PHV. It should be noted that the calculated sample power was 0.88. The margin of error pointed out for the sample size was 4.87%, indicating that the sample used in the present study has statistical strength to address the research question.

Table 1. Sample characterization.

Variables	Values
n (%)	64 (100%)
Male	47 (73%)
Female	17 (27%)
Age (years)	13.6 ± 1.17
Maturity offset (years from peak height velocity (PHV))	0.14 ± 1.82
Height (cm)	159.4 ± 11.9
Wingspan (cm)	162.1 ± 11.6
Bod mass index (kg/m^2)	20.0 ± 3.40
Body weight (kg)	51.3 ± 14.8
Fat mass (kg)	12.8 ± 6.61
Lean mass (kg)	36.9 ± 10.4
Bone mineral density (g/cm^2)	1.58 ± 0.43
Bone mineral content (g)	2.21 ± 0.84

Maturity offset was significantly and positively correlated with muscle strength of the upper limbs in all sports. Only in tennis players was a significant and positive relationship found between maturity offset and strength of the lower limbs. When controlling the effect of lean mass, maturity offset showed significant positive relationship with strength performance of the upper limbs only in jiu-jitsu, swimming, tennis and volleyball (Table 2).

The analysis of the patterns of the variables by Clusters K-means, grouped by maturity offset, were developed for hand grip strength, upper limbs strength and lean mass. The regressions from this analysis are shown in Table 3 (model 1 considers the entire group of variables, whilst model 2 removes maturity offset). Comparison between models identifies the strength of the predictions between models.

Regarding the lower limb muscle strength performance variables, the clustering occurred only for tennis and volleyball, grouping the variables vertical jump, CMJ and lean mass. Thus, the regression models were organized based on the groupings of the variables. Models 1 and 2 were significant for all sports and model 3 only for tennis and volleyball. Please note maturity offset was used as an independent variable within regression models 1 and 4. In models 2 and 3 lean mass was used as an independent variable. No multicollinearities were found among the variables within the models used.

Table 2. Correlations with somatic maturation.

Groups	Hand Grip (kgf)		ULS (m)		Vertical Jump (cm)		CMJ (cm)	
	r	p	r	p	r	p	r	p
Rowing	0.63 *	0.01	0.44	>0.05	−0.15	> 0.05	−0.18	>0.05
Swimming	0.46 *	0.01	0.13	>0.05	0.54	>0.05	0.32	>0.05
Soccer	0.68 *	0.01	0.14	>0.05	−0.11	>0.05	−0.01	>0.05
Tennis	0.72 *	<0.0001	0.76 *	0.0001	0.74 *	0.01	0.60 *	0.01
Jiu-jitsu	0.91 *	0.001	0.58	>0.05	0.57	>0.05	0.54	>0.05
Volley	0.45 *	0.01	0.08	>0.05	0.14	>0.05	0.03	>0.05
Variable Control: Lean Mass (kg)								
Rowing	0.21	>0.05	0.42	>0.05	0.60	>0.05	0.60	>0.05
Swimming	0.13	>0.05	0.58 *	0.01	0.21	>0.05	0.24	>0.05
Soccer	0.08	>0.05	0.37	>0.05	0.23	>0.05	0.20	>0.05
Tennis	0.42	>0.05	0.64 *	0.01	0.48	>0.05	0.06	>0.05
Jiu-jitsu	0.70 *	0.0001	0.16	>0.05	0.06	>0.05	0.05	>0.05
Volley	0.18	>0.05	0.73 *	0.01	0.19	>0.05	0.21	>0.05

* Statistically significant. ULS = upper limbs strength. CMJ = jump against movement.

Table 3. Linear regression adjusted to predict the strength performance of upper and lower limbs in different sports.

Groups	Model 1: Maturation/Hand Grip (kgf)/Upper Limbs Strength (m)/Lean Mass (kg)			Model 2: Lean Mass (kg)/Hand Grip (kgf)/Upper Limbs Strength (m)		
	r^2	β	p	r^2	β	p
Rowing	0.66 *	0.26	0.0001	0.90 *	3.10	<0.0001
Swimming	0.60 *	0.09	0.01	0.82 *	0.86	<0.0001
Soccer	0.63 *	0.02	0.01	0.78 *	0.95	<0.0001
Tennis	0.69 *	0.64	0.01	0.75 *	3.83	<0.0001
Jiu-jitsu	0.84 *	0.21	<0.0001	0.83 *	3.26	0.0001
Volley	0.64 *	1.52	0.01	0.91 *	3.44	<0.0001
	Model 3: Lean Mass (kg)/Vertical Jump (cm)/CMJ (cm)			**Model 4: Maturation/Vertical Jump (cm)/CMJ (cm)/Lean Mass (kg)**		
Rowing	0.21	1.48	>0.05	0.60	0.14	>0.05
Swimming	0.64	0.72	>0.05	0.63	0.07	>0.05
Soccer	0.07	1.91	>0.05	0.59	0.03	>0.05
Tennis	0.73 *	0.42	<0.0001	0.61	0.18	>0.05
Jiu-jitsu	0.53	2.52	>0.05	0.69	0.12	>0.05
Volley	0.64 *	2.90	0.01	0.22	−0.03	>0.05

* Statistically significant. CMJ = Jump against movement.

Table 4 shows different maturation stages (early, average and late) between the different sports groups (soccer and tennis before peak growth; swimming and jiu-jitsu during peak growth and rowing and volleyball after peak growth). Thus, in relation to maturation, the rowing and volleyball athletes did not point out differences between them, and both groups were the ones who indicated more advanced maturation stage in relation to the others. In addition, the rowers had a higher total lean mass than the other athletes ($p = 0.01$). In addition, rowing athletes showed superiority in body height in relation to the soccer and jiu-jitsu groups ($p = 0.01$). Furthermore, it was found that in the groups of rowing ($f^2 = 0.54$, large), soccer ($f^2 = 0.89$, large), swimming ($f^2 = 3.03$, large), jiu-jitsu ($f^2 = 0.63$, large), volleyball ($f^2 = 0.49$, large) and tennis ($f^2 = 0.33$, medium), maturation showed a large local effect in relation to lean mass. In addition, the effects of interaction of great magnitude of maturation with lean mass ($\eta^2 p = 0.753$, very large), body height ($\eta^2 p = 0.523$, very large) and body weight ($\eta^2 p = 0.631$, very large) were also verified in the analyses. The effect of the sex variable (male and female) within and between groups was not significant in the variables used in the comparisons performed by the ANCOVA test (F < 0.3; p > 0.1). Regarding maturation, the variation coefficient was 92% in rowing, 30.7% in swimming, 137% in soccer, 128% in tennis, 35.6% in jiu-jitsu and 192.8% in volley.

Table 4. Comparisons between groups regarding maturation, lean mass, height and body weight.

Groups	Maturation	Lean Mass (kg)	Stature (cm)	Body Weight (kg)
Rowing	1.54 ± 1.67 *	45.5 ± 9.88 *ᕼ	167.7 ± 9.61 *§	65.1 ± 16.7
Swimming	−0.35 ± 1.14	36.0 ± 9.75	160.6 ± 9.39	48.9 ± 10.8
Soccer	−1.76 ± 1.28	38.7 ± 9.84	157.7 ± 11.7	50.4 ± 13.2
Tennis	−1.17 ± 0.91	34.2 ± 6.45	159.0 ± 9.29	46.5 ± 7.76
Jiu-jitsu	−0.92 ± 2.58	39.7 ± 13.0	143.5 ± 49.9	47.4 ± 21.3
Volley	1.62 ± 0.84 *	36.7 ± 6.18	149.9 ± 49.9	49.7 ± 19.4

* Statistical superiority in relation to the swimming, soccer, tennis and jiu-jitsu groups for $p = 0.01$.*ᕼ Statistical superiority in relation to all groups for $p = 0.01$. *§ = Statistical superiority in relation to soccer and jiu-jitsu groups for $p = 0.01$.

When comparing muscle strength performance between groups (Figure 1), the rowers stood out in relation to upper and lower limbs strength (medicine ball: $F_{(5,0)} = 5.18$; $p = 0.01$; hand grip: $F_{(5,0)} = 4.60$; $p < 0.0001$; vertical jump: $F_{(5,0)} = 3.09$; $p < 0.0001$). In addition, maturation pointed to a large interaction with the strength of the upper limbs (HG: $\eta^2 p = 0.525$, very large; ULS: $\eta^2 p = 0.452$, very large) and small interaction with the lower limbs (VJ: $\eta^2 p = 0.140$, small; CMJ: $\eta^2 p = 0.162$, small), while the lean mass showed great interaction with the strength of upper limbs (HG: $\eta^2 p = 0.532$, very large; ULS: $\eta^2 p = 0.674$, very large) and lower (VJ: $\eta^2 p = 0.739$, very large; CMJ: $\eta^2 p = 0.735$, very large). The effect of the sex was not significant in the comparisons by the ANCOVA test (F < 1.0; $p > 0.3$).

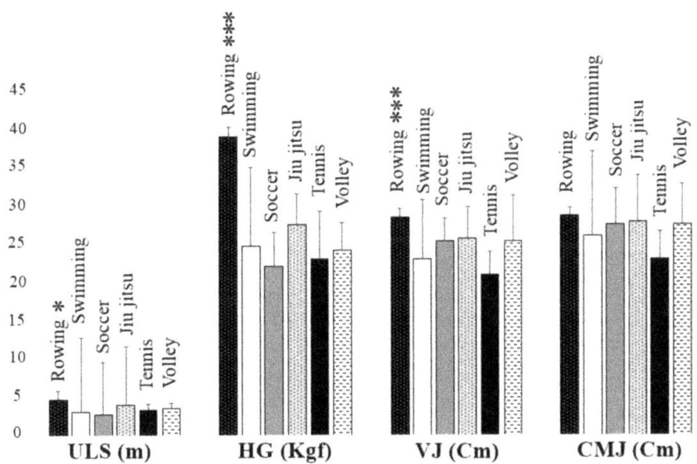

Figure 1. Comparison of muscular performance between different sports. *** $p < 0.0001$. * $p = 0.01$. ULS = upper limb strength. HG = hand grip. VJ = vertical jump. CMJ = jump against movement.

Table 5 shows the local effect of lean mass and maturation on the variables of muscle strength in the upper and lower limbs. Thus, a large effect of lean mass in relation to upper and lower limb strength of the young athletes analyzed was exposed. A similar result was found in relation to biological maturation.

Table 5. Local effect of lean mass and maturation in relation to muscle strength performance.

Groups	Hand Grip (kgf)	Upper Limbs Strength (m)	Vertical Jump (cm)	CMJ (cm)
		Lean Mass		
	f^2	f^2	f^2	f^2
Rowing	1.15	1.77	6.46	7.45
Swimming	1.59	2.84	2.99	2.93
Soccer	2.19	4.95	10.1	6.53
Tennis	3.09	3.86	3.42	3.03
Jiu-jitsu	0.64	0.53	0.08	0.34
Volley	0.03	0.50	1.46	1.01
		Maturation		
Rowing	0.18	0.19	0.16	0.14
Swimming	0.23	0.20	0.35	0.30
Soccer	0.73	5.11	4.11	4.59
Tennis	0.59	0.54	0.29	0.43
Jiu-jitsu	0.40	0.62	0.42	0.27
Volley	0.36	0.49	0.78	0.67

CMJ = jump against movement. f^2 = Cohen's coefficient, for the size of the effect of the interaction between the variances of the variables. Magnitude f^2: small: $f^2 \leq 0.02$ to 0.14; medium: $f^2 \geq 0.15$ to 0.34; large: $f^2 \geq 0.35$.

10. Discussion

The aim of this study was to analyze the independent and interaction relationships between biological maturation and lean mass on muscular strength development of the upper and lower limbs of young athletes practicing different sports. Our results showed: (i) a significant relationship between maturation and the performance of lower limbs of tennis players and upper limbs of rowers, tennis players, jiu-jitsu participants, volleyball players, swimmers and soccer players; (ii) when controlling for lean mass, a relationship between maturation and performance in the upper limbs of jiu-jitsu participants, swimmers, tennis and volleyball players was found; (iii) a relationship between lean mass and maturation, and performance of upper limbs in rowers, tennis players, jiu-jitsu participants, volleyball players, swimmers and soccer players; (iv) maturation significantly interacted with lean mass; (v) lean mass and maturation showed independent effects, in relation to the performance of upper and lower limbs; and (vi) athletes with higher concentrations of lean mass showed superior performance in the upper and lower limbs.

These findings suggest that athletes with superior lean mass present superior muscle strength in the upper and lower limbs when compared to athletes from other modalities who have less lean mass. Our results are in agreement with Azimi et al. [4] who evaluated the impact of lean mass as a predictor of strength performance in elite young Iranian weightlifting athletes. These researchers reported that lean mass showed significant relationships with specific performance and that athletes with superior lean mass were able to lift loads with higher weight compared to the others ($p \leq 0.001$).

The present study also showed that lean mass is related to muscle strength. Furthermore, in young athletes of different modalities there is a significant effect size (ES) between lean mass and ULS, HG, VJ and CMJ. Raymond-Pope et al. [14] in researching university athletes using DEXA and isokinetic dynamometry found that the lean mass of the lower limbs was related to the neuromuscular performance of jumps. The authors determined that the sum of the amount of lean mass in both legs showed strong correlations with the production of muscle strength in the lower limbs (r = 0.83–0.94; $p < 0.05$).

Another important finding in the present study was that there was a significant ES of maturation in relation to lean mass. Rowers with superior lean mass were also superior in relation to biological maturation. These results are in accord with Malina et al. [6] who showed advanced biological maturation favors the production of muscle strength, which in turn is influenced by the total volume of lean body mass. However, the results of this research also showed that the ES of maturation with the performance of muscle strength in rowers was of a lower magnitude than the ES found in other sports (see Table 5). Our results differ from previous studies, which found relationships between lean mass and advanced maturation regardless of the analyzed sport [19–22,40,41].

Moreover, it is highlighted that the specific stimuli of sports modalities can also contribute to the process of acquiring muscle strength [2,13]. Thus, according to Giroux et al. [13], rowing is a sport that requires constant traction against water resistance. In this sense, the chronic effect of the practice of specific movements to generate the drag force that promotes the displacement of the vessels may contribute to morphological adaptations in the musculoskeletal tissue, providing the mechanism for gaining lean mass [13,22].

We are in agreement with Pion et al. [26] who argued that body composition is in fact an important component for the motor skills of sports talents. In addition, we would argue that our results support the contention that sports have different ways of stimulating motor skills (i.e., training intensity, recruitment of glycolytic fibers, execution of specific movements against pre-established resistance, etc.). Concomitant with this statement, the findings of the present research indicate that lean mass has a significant ES (See Figure 1) with upper limb strength performance and that maturation has an ES with young lean rowers, soccer, tennis, volleyball and jiu-jitsu players, and swimmers.

Thus, the regression analyses carried out by the present study shows that maturation and lean mass (Table 3; model 1 and 2), combined with upper limb strength tests have the potential to predict the performance of upper limbs of rowers, swimmers, tennis, volleyball, soccer and jiu-jitsu players. For the performance of lower limb strength, the model based on lean mass and lower limb muscle strength tests indicated the ability to predict performance only in volleyball and tennis players (Table 3; model 3).

The findings of this study may help in a sustainable manner, as the experiments are not costly and advanced technology is not required for the quantitative testing, the identification of youth talent in sports programs. They should also assist in the longitudinal monitoring of young athletes and the selection of young athletes into teams and level of competition. It is important to emphasize that the results of the present study can be used for the analysis of the maturation stage, associating it with body morphology as predictors of upper and lower limb muscle strength. In addition, contributing to the direction of complementary sports training, the results suggest that the provision of stimuli that use external resistance may be useful for the acquisition of lean mass, and may contribute to the performance of muscular strength of upper and lower limbs.

However, despite the relevance of the results, the present study has some limitations: (i) The research methodology is an observational approach, which does not allow for a stabilization of cause and effect. (ii) The peculiarity of the sample is composed only of young athletes, which can make it difficult to stagger the results for young non-athletes. (iii) The control of the nutritional history of the analyzed subjects was not carried out, which makes it difficult to attribute the results related to lean mass to training stimuli or the nutritional quality of the subjects.

11. Conclusions

It is concluded that biological maturation shows a significant relationship with muscle strength, especially the upper limbs. Additionally, there is an important interaction of maturity with lean mass in this process. Regarding the local effect, the magnitude of the effect of lean mass on muscle strength is broader in relation to that of maturation, especially for the strength of lower limbs. In addition, young athletes with higher levels of lean mass have advanced maturation and superior muscle strength in the upper and lower limbs.

Author Contributions: Conceptualization, P.F.d.A.-N. and B.G.d.A.T.C.; methodology, P.F.d.A.-N., B.G.d.A.T.C., G.R.B. and P.M.S.D.; software, P.F.d.A.-N. and D.G.d.M.; validation, M.D., F.J.A., D.G.d.M. and V.C.M.P.; formal analysis, P.F.d.A.-N., M.D. and P.M.S.D.; investigation, P.F.d.A.-N., M.D., V.C.M.P., G.R.B. and B.G.d.A.T.C.; resources, V.C.M.P., F.J.A. and P.M.S.D.; data curation, P.F.d.A.-N., D.G.d.M. and F.J.A.; writing—original draft preparation, P.F.d.A.-N., D.G.d.M., A.D.G.B.-J. and B.G.d.A.T.C.; writing—review and editing, D.G.d.M., A.D.G.B.-J., G.R.B. and B.G.d.A.T.C.; visualization, P.F.d.A.-N. and B.G.d.A.T.C.; supervision, V.C.M.P. and P.M.S.D.; project administration, B.G.d.A.T.C.; funding acquisition, V.C.M.P. and B.G.d.A.T.C. All authors have read and agreed to the published version of the manuscript.

Funding: This research received no external funding.

Acknowledgments: For your support and encouragement for the development of this academic article, we thank the Federal University of Rio Grande do Norte (UFRN), the Physical Activity and Health (AFISA) research base, the Child and Adolescent Maturation Research Group (GEPMAC). The National Council for Scientific Development (CNPQ) and the Higher Education Personnel Improvement Coordination (CAPES).

Conflicts of Interest: The authors declare no conflicts of interest.

References

1. Stachon, A.; Pietraszewska, J.; Pietraszewski, B.; Andrzejewska, J.; Burdukiewicz, A. Anthropometric characteristics and lower limb power of professional female volleyball players. *S. Afr. J. Res. Sport Phys. Educ. Recreat.* **2015**, *37*, 99–112.

2. Suchomel, T.J.; Nimphius, S.; Stone, M.H. The importance of muscular strength in athletic performance. *Sports Med.* **2016**, *46*, 1419–1449. [CrossRef]

3. Matsudo, V.K.; Rivet, R.E.; Pereira, M.H. Standard score assessment on physique and performance of Brazilian athletes in a six tiered competitive sports model. *J. Sports Sci.* **1987**, *5*, 49–53. [CrossRef]

4. Azimi, F.; Siahkouhian, M.; Hedayatnejad, M. Lean body mass as a predictor of performance of young Iranian elite weightlifters. *S. Afr. J. Res. Sport Phys. Educ. Recreat.* **2016**, *38*, 179–186.

5. Torres-Unda, J.; Zarrazquin, I.; Gil, J.; Ruiz, F.; Irazusta, A.; Kortajarena, M.; Irazusta, J. Anthropometric, physiological and maturational characteristics in selected elite and non-elite male adolescent basketball players. *J. Sports Sci.* **2013**, *31*, 196–203. [CrossRef]

6. Malina, R.M.; Rogol, A.D.; Cumming, S.P.; Silva, M.J.C.; Figueiredo, A.J. Biological maturation of youth athletes: Assessment and implications. *Br. J. Sports Med.* **2015**, *49*, 852–859. [CrossRef]

7. Rommers, N.; Mostaert, M.; Goossens, L.; Vaeyens, R.; Witvrouw, E.; Lenoir, M.; D'Hondt, E. Age and maturity related differences in motor coordination among male elite youth soccer players. *J. Sports Sci.* **2019**, *37*, 196–203. [CrossRef] [PubMed]

8. Charcharis, G.; Mersmann, F.; Bohm, S.; Arampatzis, A. Morphological and Mechanical Properties of the Quadriceps Femoris Muscle-Tendon Unit from Adolescence to Adulthood: Effects of Age and Athletic Training. *Front. Physiol.* **2019**, *10*, 1082. [CrossRef] [PubMed]

9. Hunter, S.K.; Pereira, H.M.; Keenan, K.G. The aging neuromuscular system and motor performance. *J. Appl. Phys.* **2016**, *121*, 982–995. [CrossRef] [PubMed]

10. Kuhman, D.J.; Hurt, C.P. Lower extremity joints and muscle groups in the human locomotor system alter mechanical functions to meet task demand. *J. Exp. Biol.* **2019**, *222*, jeb206383. [CrossRef] [PubMed]

11. Yang, J.; Christophi, C.A.; Farioli, A.; Baur, D.M.; Moffatt, S.; Kales, S.N. Association between push-up exercise capacity and future cardiovascular events among active adult men. *JAMA Netw. Open* **2019**, *2*, e188341. [CrossRef] [PubMed]

12. Liu, J.; Yan, Y.; Xi, B.; Huang, G.; Mi, J. China Child and Adolescent Cardiovascular Health (CCACH) Study Group. Skeletal muscle reference for Chinese children and adolescents. *J. Cachexia Sarcopenia Muscle* **2019**, *10*, 155–164. [CrossRef] [PubMed]

13. Giroux, C.; Maciejewski, H.; Ben-Abdessamie, A.; Chorin, F.; Ratel, S.; Rahmani, A. Relationship between force-velocity profiles and 1.500-m ergometer performance in young rowers. *Int. J. Sports Med.* **2017**, *38*, 992–1000. [CrossRef] [PubMed]

14. Raymond-Pope, C.J.; Dengel, D.R.; Fitzgerald, J.S.; Bosch, T.A. Association of compartmental leg lean mass measured by dual X-ray absorptiometry with force production. *J. Strength Cond. Res.* **2018**. [CrossRef] [PubMed]

15. Durkalec-Michalski, K.; Nowaczyk, P.M.; Podgórski, T.; Kusy, K.; Osiński, W.; Jeszka, J. Relationship between body composition and the level of aerobic and anaerobic capacity in highly trained male rowers. *J. Sports Med. Phys. Fit.* **2019**, *59*, 1526–1535. [CrossRef]

16. Thomas, M.H.; Burns, S.P. Increasing lean mass and strength: A comparison of high frequency strength training to lower frequency strength training. *Int. J. Exerc. Sci.* **2016**, *9*, 159.

17. O'Brien, T.D.; Reeves, N.D.; Baltzopoulos, V.; Jones, D.A.; Maganaris, C.N. In vivo measurements of muscle specific tension in adults and children. *Exp. Physiol.* **2009**, *95*, 202–210. [CrossRef]

18. Cunha, G.D.S.; Vaz, M.A.; Herzog, W.; Geremia, J.M.; Leites, G.T.; Reischak-Oliveira, Á. Maturity status effects on torque and muscle architecture of young soccer players. *J. Sports Sci.* **2019**, 1–10. [CrossRef]

19. Pinto, V.C.M.; dos Santos, P.G.M.D.; Dantas, M.P.; Araújo, J.P.D.F.; Cabral, S.D.A.T.; Cabral, B.G.D.A.T. Relationship between skeletal age, hormonal markers and physical capacity in adolescents. *J. Hum. Growth Dev.* **2017**, *27*, 77–83. [CrossRef]

20. Gantois, P.; Pinto, V.; Castro, K.R.D.; João, P.V.; Dantas, P.; Cabral, B.G. Skeletal age and explosive strength in young volleyball players. *Rev. Bras. Cineantropom. Desempenho Hum.* **2017**, *19*, 331–342. [CrossRef]

21. Dantas, M.P.; Silva, L.F.; Gantois, P.; Silva, L.M.; Dantas, R.N.; Cabral, B.T. Relationship between maturation and explosive strength in young rowers. *Motricidade* **2018**, *14*, 112–121.

22. Hammami, M.; Hermassi, S.; Gaamouri, N.; Aloui, G.; Comfort, P.; Shephard, R.J.; Chelly, M.S. Field tests of performance and their relationship to age and anthropometric parameters in adolescent handball players. *Front. Physiol.* **2019**, *10*, 1124. [CrossRef] [PubMed]

23. De Almeida-Neto, P.F.; Silva Dantas, P.M.; Pinto, V.C.M.; Cesário, T.D.M.; Ribeiro Campos, N.M.; Santana, E.E.; Matos, D.G.; Aidar, F.J.; Cabral, B.G.D.A.T. BiologicalMaturationand Hormonal Markers, Relationshipto Neuromotor Performance in FemaleChildren. *Int. J. Environ. Res. Public Health* **2020**, *17*, 3277. [CrossRef]

24. Mujika, I.; Rønnestad, B.R.; Martin, D.T. Effects of increased muscle strength and muscle mass on endurance-cycling performance. *Int. J. Sports Phys. Perf.* **2016**, *11*, 283–289. [CrossRef]

25. Prieske, O.; Muehlbauer, T.; Granacher, U. The role of trunk muscle strength for physical fitness and athletic performance in trained individuals: A systematic review and meta-analysis. *Sports Med.* **2016**, *46*, 401–419. [CrossRef]

26. Pion, J.; Hohmann, A.; Liu, T.; Lenoir, M.; Segers, V. Predictive models reduce talent development costs in female gymnastics. *J. Sports Sci.* **2017**, *35*, 806–811. [CrossRef]

27. STROBE Checklist Strengthening the Reporting of Observational Studies in Epidemiology. Available online: https://www.strobe-statement.org/index.php?id=strobe-home (accessed on 15 June 2020).

28. Karupaiah, T. Limited (ISAK) profiling the International Society for the Advancement of Kinanthropometry (ISAK). *J. Ren. Nutr. Metab.* **2018**, *3*, 11. [CrossRef]

29. Mirwald, R.L.; Baxter-Jones, A.D.; Bailey, D.A.; Beunen, G.P. An assessment of maturity from anthropometric measurements. *Med. Sci. Sports Exerc.* **2002**, *34*, 689–694.

30. Khadilkar, A.; Chiplonkar, S.; Sanwalka, N.; Khadilkar, V.; Mandlik, R.; Ekbote, V. A Cross-Calibration Study of GE Lunar iDEXA and GE Lunar DPX Pro for Body Composition Measurements in Children and Adults. *J. Clin. Densitom.* **2019**. [CrossRef]

31. Mello, J.B.; Nagorny, G.A.K.; Haiachi, M.D.C.; Gaya, A.R.; Gaya, A.C.A. ProjetoEsporteBrasil: Physical fitness profile related to sport performance of children and adolescents. *Rev. Bras. Cineantropom. Desempenho Hum.* **2016**, *18*, 658–666. [CrossRef]

32. Reijnierse, E.M.; de Jong, N.; Trappenburg, M.C.; Blauw, G.J.; Butler-Browne, G.; Gapeyeva, H.; Stenroth, L. Assessment of maximal handgrip strength: How many attempts are needed? *J. Cachexia Sarcopenia Muscle* **2017**, *8*, 466–474. [CrossRef] [PubMed]

33. Forza, J.; Edmundson, C.J. Comparison between Gyko inertial sensor and Chrono jump contact mat for the assessment of Squat Jump, Countermovement Jump and Abalakov Jump in amateur male volleyball players, amateur male rugby players and in high school students. *J. Multidiscip. Eng. Sci. Technol.* **2019**, *6*, 9982–9988.

34. Schober, P.; Boer, C.; Schwarte, L.A. Correlation coefficients: Appropriate use and interpretation. *Anesth. Anal.* **2018**, *126*, 1763–1768. [CrossRef] [PubMed]

35. Hartigan, J.A.; Wong, M.A. Algorithm AS 136: A k-means clustering algorithm. *J. R. Stat. Soc. Ser. C (Appl. Stat.)* **1979**, *28*, 100–108. [CrossRef]

36. Pearl, J. *Causality*; Cambridge University Press: Cambridge, UK, 2009; ISBN 978-0521895606.

37. Lee, P.H. Should we adjust for a confounder if empirical and theoretical criteria yield contradictory results? A simulation study. *Sci. Rep.* **2014**, *4*, 6085–6099. [CrossRef]

38. Landau, S.; Everitt, B.S. *A Handbook of Statistical Analyses Using SPSS*; Chapman—Hall/CRC: Washington, DC, USA, 2004; ISBN 1-58488-369-3.

39. Cohen, J. *Statistical Power Analysis for the Behavioral Sciences*, 2nd ed.; Lawrence Erlbaum Associates: Hillsdale, NJ, USA, 1988.

40. Scheffler, C.; Hermanussen, M. Growth in childhood and adolescence. *Int. Encycl. Biol. Anthropol.* **2018**. [CrossRef]
41. Almeida-Neto, P.F.; Matos, D.G.; Pinto, V.C.M.; Dantas, P.M.S.; Cesário, T.M.; da Silva, L.F.; Bulhões-Correia, A.; Aidar, J.F.; Cabral, B.G.A.T. Can the Neuromuscular Performance of Young Athletes Be Influenced by Hormone Levels and Different Stages of Puberty? *Int. J. Environ. Res. Public Health* **2020**, *17*, 5637. [CrossRef]

Article

Effects of Resistance Training with Different Pyramid Systems on Bioimpedance Vector Patterns, Body Composition, and Cellular Health in Older Women: A Randomized Controlled Trial

Leandro dos Santos [1], Alex S. Ribeiro [2], Luís A. Gobbo [3], João Pedro Nunes [1,*], Paolo M. Cunha [1], Francesco Campa [4], Stefania Toselli [5], Brad J. Schoenfeld [6], Luís B. Sardinha [7] and Edilson S. Cyrino [1]

[1] Metabolism, Nutrition, and Exercise Laboratory, Physical Education and Sports Center, Londrina State University, Londrina, PR 86057-970, Brazil; leandro.santos.sm@gmail.com (L.d.S.); pcunha88m@gmail.com (P.M.C.); edilsoncyrino@gmail.com (E.S.C.)

[2] Center for Research in Health Sciences, University of Northern Paraná, Londrina, PR 86041-140, Brazil; alex.sribeiro@kroton.com.br

[3] Department of Physical Education, São Paulo State University, Presidente Prudente, SP 19060-900, Brazil; luis.gobbo@unesp.br

[4] Department for Life Quality Studies, University of Bologna, 47921 Rimini, Italy; francesco.campa3@unibo.it

[5] Department of Biomedical and Neuromotor Science, University of Bologna, 40126 Bologna, Italy; stefania.toselli@unibo.it

[6] Exercise Science Department, CUNY Lehman College, Bronx, NY 10468, USA; bradschoenfeldphd@gmail.com

[7] Exercise and Health Laboratory, CIPER, Faculdade de Motricidade Humana, Universidade de Lisboa, 1499-002 Lisboa, Portugal; lbsardinha55@gmail.com

* Correspondence: joaonunes.jpn@hotmail.com

Received: 20 July 2020; Accepted: 16 August 2020; Published: 18 August 2020

Abstract: Bioelectrical impedance vector analysis (BIVA) and phase angle (PhA) have been widely used to monitor changes in health-related parameters in older adults, while resistance training (RT) is one of the potential strategies to mitigate the adverse effects of aging. The purpose of this study was to compare the effects of the crescent pyramid RT system with two repetition zones on BIVA patterns and PhA. Fifty-five older women ($\geq$60 years) were randomly assigned into three groups: control (CON, n = 18), narrow pyramid (NPR, n = 19), and wide pyramid (WPR, n = 18). The RT was performed for eight weeks, three times per week, in eight exercises for the whole body with three sets of 12/10/8 (NPR) or 15/10/5 repetitions (WPR). Bioimpedance spectroscopy (50 kHz frequency) was assessed. After the intervention period, both training groups showed significant changes in BIVA patterns compared to CON ($p < 0.001$); resistance decreased and reactance increased, which resulted in a BIVA-vector displacement over time ($p < 0.001$). Changes in PhA were greater for WPR ($\Delta\%$ = 10.6; effect size [ES] = 0.64) compared to NPR ($\Delta\%$ = 5.3; ES = 0.41) and CON ($\Delta\%$ = −6.4; ES = −0.40). The results suggest that the crescent pyramid RT system with both repetition zones (WPR and NPR) is effective for inducing improvements in BIVA patterns and PhA in older women, although WPR elicits greater increases in PhA than NPR.

Keywords: aging; BIVA; bioelectrical impedance analysis; body composition; elderly; strength training; dose-response

---

1. Introduction

Aging is associated with a variety of adverse changes to older adults [1–4]. Sarcopenia, described as the loss of muscle mass and function, is one of the main health problems in aging and has been associated with increased risk of falls, injuries, and hospitalizations [1]. The cost of care and treatment for the elderly who suffer fatal and nonfatal falls is high and burdens health systems [1,2]. Therefore, implementing strategies to improve functionality and body composition in older adults such as physical exercises are deemed of primary importance to lead a healthy life [1,3], and may mitigate socioeconomic impacts by reducing health care costs [1,2]. In this way, analyzing and monitoring body composition is an essential topic when discussing the benefits of leading an active lifestyle.

Among the several tools to estimate body composition, the bioelectrical impedance analysis (BIA) is a non-invasive, inexpensive, and simple method to do it in different populations [5]. Recent studies have opted to use the interpretation of the raw parameters obtained from BIA, like bioelectrical impedance vector analysis (BIVA) and phase angle (PhA), to infer regarding body composition, instead of using prediction equations to avoid errors associated with them [6]. Through the BIVA approach, the bioelectric parameters of resistance (R) and reactance (Xc) are interpreted together as a vector within an R-Xc graph. Since R and Xc reflect body fluid content and cell density, respectively, the position of the bioimpedance vector provides essential information on body composition [5]. Additionally, the vector inclination determines the bioelectrical PhA, which represents the relationship between the intra- (ICW) and extracellular fluids (ECW) [7].

The aging-associated alterations in body composition, manifested by reduction of skeletal muscle mass and increased body fat, may influence BIVA parameters resulting in PhA decreases [8–12]. In contrast, exercise practice promotes positive changes in BIVA and PhA, followed by benefits on psychological, cognitive, and physical aspects, hydration, nutritional status, muscle function, and quality of life in the elderly [11–20]. Resistance training (RT) is one of the potential strategies to reverse the adverse effects of aging on cellular integrity and function, improve BIA parameters, and induce changes in the cellular volume of skeletal muscle tissue and cell membrane potential [8–13]. However, despite some recent and preliminary results [11,21–23], the dose–response relationship regarding specific RT variables (e.g., training period, frequency, intensity, volume, training system) on bioelectrical parameters is unclear.

In this regard, the pyramid training system is an RT strategy applied to increase muscular strength and muscle mass, whereby loads are progressively increased (crescent pyramid) or decreased (reverse pyramid) inversely to the number of repetitions for successive sets of a given exercise [24,25]. The traditional crescent pyramid system adopts a narrow repetition zone (e.g., 12, 10, and 8 repetitions, respectively) with ~5% load increases for each set [24]. However, due to the muscular strength to be reduced with the aging process, older adults have shown difficulty in increasing the exercise loads throughout the sets in the proper manner [24]. Thus, both traditional and crescent pyramid systems tend to result in similar adaptative responses to RT in older adults [24]. Therefore, it seems that an alternative for this population can be the use of a wider repetition zone such as 15, 10, and 5 repetitions, respectively, for each set [25]. From a wider repetition zone, it would be possible for progressive increases of the weight to allow an appropriate combination of metabolic and mechanic stimuli, which might result in additional benefits. This hypothesis was confirmed in a recent study that found a better hypertrophic response in older women submitted to a crescent pyramid RT system with a wider variety of training loads than a narrower loading range [25].

Therefore, considering the correlation between muscular strength and muscle mass with PhA in older men and women [13,26], it seems logical that a crescent pyramid RT system using a more extensive repetition zone would also induce more significant changes in both BIVA patterns and PhA. Thus, the purpose of the present study was to compare the effects of the crescent pyramid system performed with a wide versus a narrow repetition zone on the most informative bioimpedance parameters in older women. It was hypothesized that training would be adequate to induce changes in BIVA vector displacements and increases in PhA and that the crescent pyramid system performed in a wide repetition zone would elicit more significant improvements compared to the narrow zone.

2. Materials and Methods

2.1. Experimental Design

The present study is part of a longitudinal research project named "Active Aging Longitudinal Study", initiated in 2012, whose purpose is to analyze the effects of supervised, structured, and progressive RT program on neuromuscular, morphological, physiological, and metabolic outcomes in older women [25]. A randomized controlled trial was carried out over 12 weeks, with eight weeks dedicated to the RT program, and four weeks for data collection. Pre- and post-intervention testing was carried out at weeks 1–2 and 11–12, respectively. The RT program was carried out during weeks 3–10. Adherence to the RT program was established as >85% of the total sessions. Participants were instructed not to perform any other type of physical exercise throughout the study period. The procedures were conducted according to the Declaration of Helsinki, and the Londrina State University Ethics Committee approved this investigation (committee opinion number: 1.306.507). No adverse event occurred during the intervention period.

2.2. Subjects

Recruitment was carried out through the newspaper, television programs, radio advertisings, and home delivery of leaflets in the central area and residential neighborhoods. Interested individuals completed detailed health history and physical activity questionnaires. Participants were subsequently admitted to the study if they met specific inclusion criteria: female, $\geq$60 years old, physically independent, free from cardiac dysfunction, not receiving hormonal replacement therapy, and not performing any regular physical exercise for more than once a week over the six months preceding the beginning of the study. Participants passed a diagnostic test by a cardiologist (resting 12-lead electrocardiogram test, personal interview, and treadmill stress test when deemed necessary). All were released with no restrictions for participation in this study. Fifty-nine physically independent older women (67.3 ± 4.4 years, 66.5 ± 12.6 kg, 1.55 ± 0.1 m, 27.6 ± 5.0 kg·m^{-2}) were selected and randomly assigned to one of three groups: nonexercise control group (CON, n = 19); pyramid RT system with narrow repetition zone (NPR, n = 20), in which participants performed three sets of 12/10/8 repetitions per exercise, respectively; or pyramid RT system with wider repetition zone (WPR, n = 20), in which participants performed three sets of 15/10/5 repetitions per exercise, respectively. This final number of subjects reached the necessary criteria for this experiment, according to the sample size calculation (repeated measures, moderate effect size = 0.50, α = 0.05, power = 0.80). Written informed consent was obtained from all participants after a detailed description of study procedures was provided.

2.3. Bioimpedance Spectroscopy

A phase-sensitive BIA (Xitron Hydra, model 4200, Xitron Technologies, San Diego, CA, USA) was used to obtain whole-body R and Xc at a single frequency at 50 kHz. PhA was calculated as arc-tangent (Xc/R) × 180°/π. Classic BIVA values were calculated relative to height (R/H and Xc/H). Fat-free mass (FFM), fat mass (FM), total body water (TBW), and its fractions ECW and ICW, were assessed by equations on the software device. Before each test, the analyzer was calibrated by measuring, modeling, and computing volume using a module provided by the manufacturer. The calibration test result is based on the default ECW and ICW resistivity coefficients. Participants were instructed to lie in a supine position for approximately 10 min (serving as an equilibration period). After cleaning the skin with alcohol, four electrodes were positioned on the surface of the right hand and right foot, according to conventional procedures established in the literature [27]. Participants were instructed to urinate about 30 min before the measures, refrain from ingesting food or drink in the last four hours, avoid strenuous physical exercise for at least 24 h, refrain from the consumption of alcoholic and caffeinated beverages for at least 48 h, and avoid the use of diuretics at least during seven days prior to each assessment. The BIA device was calibrated each day according to the manufacturer's recommendations. The exams were performed by the same professional in the pre- and post-intervention periods. The intraclass correlation coefficient (ICC) and standard error of measurement (SEM) were = [ECW: (SEM = 0.32 L, ICC = 0.98), ICW: (SEM = 0.19 L, ICC = 0.99), TBW: (SEM = 0.38 L, ICC = 0.98), R: (SEM = 15.6 ohms, ICC = 0.95), Xc: (SEM = 3.5 ohms, ICC = 0.96), PhA: (SEM = 0.21 degrees, ICC = 0.96)].

2.4. Resistance Training Program

A schematic representation of how RT was prescribed is displayed in Table 1. The training program was performed in the morning hours on Mondays, Wednesdays, and Fridays for eight weeks. It was based on recommendations for RT in an older population to improve muscle hypertrophy and muscular strength [28]. Physical Education professionals personally supervised (1–2 supervisors per exercise) all training sessions to reduce deviations from the study protocol and ensure safety. The RT protocol consisted of eight exercises for the whole body (as shown in Table 1), performed in either three sets of 12/10/8 RM (NPR) or 15/10/5 RM (WPR) with incrementally higher loads for each set (crescent pyramid system) [24,25]. The supervisors adjusted exercise loads according to the participant's ability and improvements in exercise capacity throughout the study to ensure that they used adequate resistance while maintaining proper technique.

Table 1. Resistance training program performed by the older women for eight weeks.

	Narrow Repetition Zone Training	Wide Repetition Zone Training
Exercises performed	chest press, horizontal leg press, seated low-row, leg extension, barbell preacher curl, lying leg curl, triceps pushdown, seated calf raise	
Number of sets × repetitions	3 × 12/10/8	3 × 15/10/5
Intensity	12/10/8 RM-load	15/10/5 RM-load
Load progression	2–5% and 5–10% per week for upper- and lower-body exercises, respectively	
Execution velocity	1 s and 2 s for concentric and eccentric movement phases, respectively	
Rest intervals	1–2 min and 2–3 min between sets and exercises, respectively	

Note. RM = repetitions-maximum.

The loads and the number of repetitions performed during each set of the eight exercises were individually recorded for each training session. The volume for each set of all exercises was calculated by multiplying the load by the number of repetitions. The volume of each exercise per session was calculated as the sum of the volume of all three sets for each exercise. The total volume-load per session was calculated as the sum of all eight exercises. The weekly volume-load (WVL) was calculated by summing the three training sessions performed in one week. Increases in WVL throughout the RT program were calculated as the WVL of the eighth week minus the WVL of the first week.

2.5. Dietary Intake

Food intake was assessed by the 24-h dietary recall method applied on two nonconsecutive days of the week, with the aid of a photographic record taken during an interview. Dietary intake was monitored in the first two and the last two weeks of the intervention period. The homemade measurements of the nutritional values of food were converted into grams and milliliters by the online software Virtual Nutri Plus (Keeple®, Rio de Janeiro, RJ, Brazil) for diet analysis. Some foods were not found in the program database, and therefore these items were added from food tables.

2.6. Statistical Analyses

The Shapiro–Wilk test was used to analyze the distribution of data. Generalized estimated equations (GEE) analyses were applied to investigate the effects of intervention over time within and between groups. Bonferroni post hoc test was adopted when significant effects on group, time, or interaction were confirmed. The paired one-sample Hotelling T^2-test [29] was used to evaluate if the changes in the mean group vectors (measured before and after the intervention period) were significantly different from zero (null vector); a 95% confidence ellipse excluding the null vector indicated a significant vector displacement. Cohen's effect size (ES) was calculated as post-training mean minus pre-training mean divided by the pooled pre-training standard deviation [30]. An ES of 0.00–0.19 was considered as trivial, 0.20–0.49 was considered as small, 0.50–0.79 as moderate, and ≥0.80 as large [30]. For all statistical analyses, statistical significance was established at $p < 0.05$. The data were stored and analyzed using IBM SPSS Statistics, v. 22.0 (IBM Corp., Armonk, NY, USA).

3. Results

Data from 55 participants were considered for final analysis (CON = 18, NPR = 19, and WPR = 18). Sample losses were due to personal reasons (CON = 1) and adherence lower than 85% (NPR = 1 and WPR = 2). A significantly greater increase in WVL was observed ($p < 0.001$) for WPR compared to NPR (+135.2 ± 15.7 kg versus +123.6 ± 17.5 kg, respectively). There were no significant differences ($p > 0.05$) in daily relative energy and macronutrients intra- and inter-groups over time (Table 2).

Table 2. Dietary intake at pre- and post-intervention according to the group.

		Control (n = 18)	Narrow Repetition Zone (n = 19)	Wide Repetition Zone (n = 18)	Interaction p-Value
Carbohydrate (g·kg·d^{-1})	Pre	3.4 ± 1.1	3.0 ± 1.0	3.0 ± 1.0	
	Post	3.4 ± 1.3	3.0 ± 1.0	3.1 ± 0.9	0.548
Protein (g·kg·d^{-1})	Pre	1.1 ± 0.3	0.9 ± 0.4	1.0 ± 0.2	
	Post	1.0 ± 0.3	1.0 ± 0.4	0.9 ± 0.2	0.091
Lipid (g·kg·d^{-1})	Pre	0.7 ± 0.2	0.7 ± 0.3	0.7 ± 0.2	
	Post	0.7 ± 0.2	0.6 ± 0.3	0.7 ± 0.2	0.632
Energy (kcal·kg·d^{-1})	Pre	25.4 ± 7.8	21.4 ± 7.6	21.9 ± 5.7	
	Post	24.3 ± 6.9	20.9 ± 8.0	22.2 ± 5.5	0.376

Note. Data are expressed as mean ± standard deviation.

Table 3 depicts the baseline and post-intervention scores, percentage changes and effect-sizes for body composition, and BIA variables by group. Both training groups had significantly ($p < 0.05$) greater increases compared to CON for FM (NPR = −0.5 kg; WPR = −0.6 kg; CON = +0.2 kg), FFM (NPR = +0.5 kg; WPR = +0.7 kg; CON = −0.2 kg), ECW (NPR = −0.3 L; WPR = −0.7 L; CON = +0.2 L), ICW (NPR = +0.7 L; WPR = +1.2 L; CON = −0.8 L), R (NPR = −8.6 ohm; WPR = −25.3 ohm; CON = +15.1 ohm), R/H (NPR = −5.4 ohm/m; WPR = −16.5 ohm/m; CON = +9.8 ohm/m), Xc/H (NPR = +1.3 ohm/m; WPR = +2.2 ohm/m; CON = −1.6 ohm/m), and Z/H (NPR = −5.8 ohm/m; WPR = −16.4 ohm/m; CON = +9.8 ohm/m) with no statistically significant differences between experimental groups ($p > 0.05$).

Figure 1 depicts the pre- to post-intervention percentage change on PhA by the group. There was a significantly higher post-study increase in PhA ($p < 0.05$) in WPR compared to NPR and CON (+10.6%, 5.3%, and −6.4%, respectively). The mean differences in R/H and Xc/H vectors with 95% confidence ellipses by groups are depicted in Figure 2. A significant change was observed for CON ($T^2 = 42.0$), NPR ($T^2 = 29.4$), and WPR ($T^2 = 57.9$), in which the 95% confidence ellipses did not cross the origin. Figure 3 presents the mean impedance vectors with 95% confidence ellipses baseline to post-intervention by groups. The vector change was statistically significant only in the WPR group ($T^2 = 8.1$; $p = 0.03$).

Table 3. Participant's scores at pre- and post-intervention according to the groups.

		Control (n = 18)	Narrow Repetition Zone (n = 19)	Wide Repetition Zone (n = 18)	Interaction *p*-Value
Fat mass (kg)	Pre	25.8 ± 9.8	27.4 ± 8.9	28.7 ± 8.0	<0.001
	Post	26.0 ± 10.0 *	26.9 ± 9.1 *	28.1 ± 8.1 *	
	Δ%	+0.77	−1.82	−2.09	
	ES	0.02	−0.05	−0.07	
Fat-free mass (kg)	Pre	38.1 ± 5.4	37.6 ± 4.7	37.5 ± 4.7	<0.001
	Post	37.9 ± 5.4 *	38.1 ± 4.8 *	38.2 ± 4.6 *	
	Δ%	−0.52	1.33	1.05	
	ES	−0.04	0.11	0.09	
TBW (L)	Pre	28.9 ± 5.4	29.3 ± 3.7	28.0 ± 3.6	0.029
	Post	28.5 ± 5.1	29.6 ± 3.6	28.4 ± 3.6 *	
	Δ%	−1.38	1.02	1.43	
	ES	−0.07	0.08	0.11	
ECW (L)	Pre	13.2 ± 2.0	12.9 ± 1.4	13.2 ± 1.6	<0.001
	Post	13.4 ± 2.0 *	12.6 ± 1.2 *	12.5 ± 1.7 *	
	Δ%	1.51	−2.33	−5.30	
	ES	0.10	−0.21	−0.44	
ICW (L)	Pre	15.6 ± 3.5	16.4 ± 2.5	14.7 ± 2.3	<0.001
	Post	14.8 ± 2.7 *	17.1 ± 2.6 *	15.9 ± 2.2 *	
	Δ%	−5.13	4.27	8.16	
	ES	−0.23	0.28	0.52	
Resistance (ohm)	Pre	572.1 ± 72.3	564.0 ± 44.8	574.5 ± 55.8	0.007
	Post	587.2 ± 75.7 *	555.4 ± 51.2 *	549.2 ± 60.5 *	
	Δ%	2.64	−1.52	−4.40	
	ES	0.21	−0.19	−0.45	
Reactance (ohm)	Pre	54.7 ± 5.0	53.6 ± 5.1	54.5 ± 6.8	0.240
	Post	53.6 ± 7.0	55.7 ± 5.9	57.8 ± 5.4 *	
	Δ%	−2.01	3.92	6.05	
	ES	−0.22	0.41	0.48	
Phase angle (degree)	Pre	5.61 ± 0.9	5.46 ± 0.7	5.48 ± 0.9	<0.001
	Post	5.25 ± 0.6	5.75 ± 0.6 *	6.06 ± 0.7 *	
	Δ%	−6.42	5.31	10.60	
	ES	−0.40	0.41	0.64	
R/H (ohm/m)	Pre	366.3 ± 40.1	366.7 ± 23.4	367.4 ± 32.0	<0.001
	Post	376.1 ± 43.1 *	361.3 ± 28.6 *	350.9 ± 35.9 *	
	Δ%	2.68	−1.47	−4.49	
	ES	0.24	−0.23	−0.51	
Xc/H (ohm/m)	Pre	36.0 ± 4.2	34.9 ± 3.1	35.0 ± 3.1	<0.001
	Post	34.4 ± 4.4 *	36.2 ± 3.7 *	37.2 ± 3.6 *	
	Δ%	−4.44	3.72	6.29	
	ES	−0.38	0.42	0.71	
Z/H (ohm/m)	Pre	366.4 ± 9.2	366.8 ± 5.2	367.4 ± 7.3	<0.001
	Post	376.2 ± 9.9 *	361.4 ± 6.3 *	351.0 ± 8.2 *	
	Δ%	2.67	−1.47	−4.46	
	ES	1.06	−1.03	−2.19	

Notes. TBW: total body water; ECW: extracellular water; ICW: intracellular water; R/H: resistance by height; Xc/H: reactance by height; Z/H: impedance by height; ES: effect size. Data are expressed as mean ± standard deviation. * $p < 0.05$ versus pre-intervention.

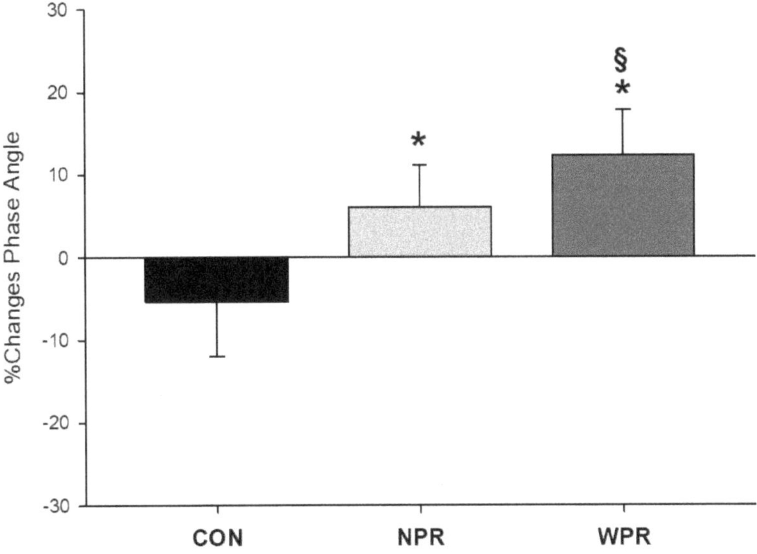

Figure 1. Percentage changes on phase angle after intervention for control (CON, n = 18), narrow pyramid (NPR, n = 19), and wide pyramid (WPR, n = 18) groups. * $p < 0.05$ versus CON. § $p < 0.05$ versus NPR.

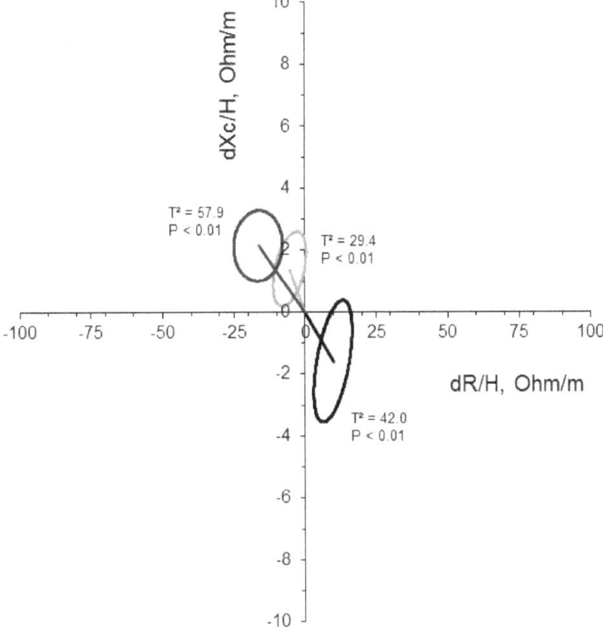

Figure 2. Mean difference vector displacement with the 95% confidence ellipses of the differences for control (black), narrow pyramid (light grey), and wide pyramid (dark grey) groups.

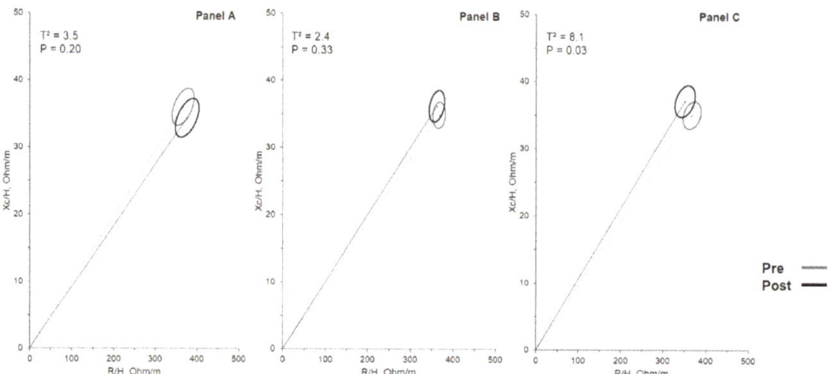

Figure 3. Mean impedance vectors with 95% confidence ellipses for control (**Panel A**), narrow pyramid (**Panel B**), and wide pyramid (**Panel C**) groups at pre- and post-intervention.

4. Discussion

To our knowledge, this is the first investigation comparing the effects of different RT systems on BIA and BIVA parameters in older women. The main finding of the current study was that a crescent-pyramid system performed with narrow or wide repetition zones was effective to improve markers of cellular integrity and function as assessed by changes in BIA outcomes. In particular, the vector displacements measured in the two intervention groups have shown how exercise has counteracted the aging effects while preserving muscle mass and cellular health. On the contrary, in the control group, the results presented an opposite trend where R increased and Xc decreased, indicating a reduction in fluid content and cellular density, respectively. As a consequence of these physiological changes and the effects of exercise, PhA increased in the training groups and decreased in the control group after the 8-week intervention period. In fact, in parallel with the bioimpedance changes, an increase in FFM was measured in the training groups versus a reduction in the control group. Also, the magnitude of the effect was influenced by the width of the repetition zone, with a wider zone inducing superior changes. Therefore, our hypothesis that RT performed with WPR would result in better improvements was confirmed.

The results of the present study indicate that performing a crescent PR system with a wider range of repetitions may maximize the metabolic response in the initial sets and the mechanical effects in the latter sets, thereby heightening anabolism [31]. Our results may be, at least in part, explained by the higher WVL verified for WPR. RT-associated muscle hypertrophy appears to be volume-dependent [32,33]. Therefore, due to its relation with muscle hypertrophy, it is possible to believe that a greater WVL elicits improvement in the BIA and BIVA parameters.

Fukuda et al. [8] reported that RT improved BIVA parameters in older women. Our findings revealed a similar change in the BIVA vectors in both training groups, indicating that increases were not influenced by different schemes of the pyramid RT system. However, for both CON and NPR groups, the R/H and Xc/H components contributed equally to the BIVA vector displacement. In contrast, for the WPR group, the change occurred due to the more significant reduction of the R/H component. Also, RT performed in the WPR system resulted in superior effect-sizes in Xc/H and ICW (ES = 0.71 and 0.52, respectively) compared to NPR (ES = 0.42 and 0.28, respectively) and Z/H (ES = 2.19 versus 1.03, respectively). These changes indicate a possible improvement in the integrity of the cell membrane, increasing the capacitance of the cells [7,12]. It has been postulated that cellular swelling in response to RT may induce an increased pressure on the cell membrane, which, in turn, threatens the cell's integrity and thus leads to increased anabolic signaling and processes to enhance the cell's ultrastructure [31].

While improvements in cell integrity, structure, and function may reflect changes in PhA, aging per se results in losses in cellular integrity and functions [8,12], reflecting reductions in PhA values, as verified in the present study (−6.4%). The PhA changes highlighted in this study are in line with the results of other studies in which the PhA increased in the intervention groups and decreased in the older women who were not involved in the physical activity program [9,10]. In healthy older women, the reference values for PhA are between 5 and 6 degrees [34]. The PhA of our sample was within the population mean in all groups (CON = 5.61 degrees, NPR = 5.46 degrees, and WPR = 5.48 degrees). Besides, our analysis revealed that in both training groups, the increases in PhA were significantly higher than the CON group. However, the rise in WPR was statistically superior to NPR ($p < 0.05$; ES = 0.64 versus 0.41, respectively), indicating that performing a wider repetition range may promote additional effects on cellularity, cell size, and integrity of cell membrane. It is not clear whether alterations in metabolic stress may have contributed to the differences found in PhA between groups; therefore, further studies are needed to determine potential explanatory mechanisms.

The alterations found for PhA of the pre- to post-intervention were higher than SEM from our lab in 83% for CON and WPR, and 89% for NPR. Our results are consistent with previous studies from our group in other cohorts of older women, which show that RT interventions lasting 8 to 12 weeks result in increments between 0.2 and 0.4 degrees (+3% to +17%) [10,11,13,21,22]. The analysis of vector displacement in each group revealed that there were increments relating to cellularity and soft tissue only in the RT groups, while in the CON group, changes were towards cachexia, fluid increase, and soft tissue reduction (Figure 2). Complementary analysis of the pre- to post-intervention vector difference showed significant differences only in the NPR group ($T^2 = 8.1$, $p = 0.03$), with a bias in favor of increased cell membrane integrity and cellularity. A clinical effect was evident (ES = 2.19) in the Z/H analysis (Table 3).

Although RT is an intervention frequently used to improve the health of women, further research is needed to reveal optimal dose–response relationships following RT in healthy older women [28]. In this regard, our results contribute significantly to practitioners, along with trainers and exercise professionals who work with older women, aiding in more scientific evidence-based exercise conduct. Different RT systems over the training mesocycles may be used as a strategy to avoid a plateau of adaptations, to increase the motivation, or still to reduce the monotony of the training sessions. Thus, the use of the crescent pyramid RT system with a wide repetition zone seems to be a feasible alternative and more appropriate for older women than the traditional pyramid system with a narrow repetition zone since it allows a higher load progression and, consequently, a better combination of metabolic and mechanic stimuli. Therefore, the crescent pyramid RT system with a wide repetition zone may promote additional benefits associated with increasing of PhA, which ultimately can mitigate the risks of falls, fractures, and reduce the number of hospitalizations, enabling greater sustainability of public health systems.

Our study has some limitations that should be addressed. First, our findings are applicable to single-frequency bioimpedance equipment. In fact, different results in measuring R and Xc values are obtained using devices that work on single- or multi-frequency [35]. Additionally, eight weeks of intervention can be considered as a relatively short period. Therefore, it is necessary to determine whether results would differ over a longer timeframe. Another aspect concerns the sample size, which included untrained older women. Thus, results should not be generalized to other populations with different age and physical fitness, as various adaptative responses to RT may be influenced by such factors. Moreover, the absence of biomarkers of lipid peroxidation, protein oxidation, and metabolic stress limit the understanding of possible mechanisms associated with the observed results. Finally, physical activity and sedentary behavior were not monitored during the experiment, hindering our ability to determine whether these factors influenced changes found in this investigation. On the other hand, it is essential to highlight the strengths of our study. All training sessions were supervised by professionals with RT experience to ensure participant safety, quality of execution of the movement, and effectiveness. The importance of supervised RT in older adults has been reported in several studies [36]. Moreover, the load adjustments were continuous and based on the participants' progress throughout the RT sessions, which permitted the maintenance of the intensity throughout the intervention. In addition, the monitoring of food consumption of the participants allowed a more consistent analysis of RT effects.

5. Conclusions

The impact of physical activity and healthy habits on body composition are particularly important in the elderly. Our findings suggest that the crescent PR system performed with narrow or wide repetition zones is effective in improving cellular integrity, function, and health (BIVA patterns and PhA). However, while the beneficial effect of physical activity in contrasting the aging process has been confirmed, WPR seems to promote superior increases in PhA in untrained older women.

Author Contributions: Conceptualization, L.d.S., A.S.R., L.A.G., J.P.N., P.M.C., F.C., S.T., B.J.S., L.B.S. and E.S.C.; data curation, L.d.S.; formal analysis, L.d.S. and L.A.G.; funding acquisition, L.B.S. and E.S.C.; investigation, L.d.S., A.S.R., J.P.N. and P.M.S.; methodology, L.d.S., A.S.R., L.B.S. and E.S.C.; project administration, L.d.S., A.S.R., J.P.N. and P.M.S.; resources, E.S.C.; supervision, L.d.S., A.S.R., J.P.N. and P.M.S.; writing—original draft, L.d.S.; writing—review and editing, L.d.S., A.S.R., L.A.G., J.P.N., P.M.S., F.C., S.T., B.J.S., L.B.S. and E.S.C. All authors have read and agreed to the published version of the manuscript.

Funding: This research received no external funding.

Acknowledgments: The authors would like to express thanks to all the participants for their engagement in this study, the Coordination of Improvement of Higher Education Personnel (CAPES/Brazil) for the scholarship conferred to J.P.N. (master), L.d.S., and P.M.C. (doctoral), and the National Council of Technological and Scientific Development (CNPq/Brazil) for the grants conceded to E.S.C. This study was partially supported by the Ministry of Education (MEC/Brazil) and CNPq/Brazil.

Conflicts of Interest: The authors declare no conflict of interest.

References

1. Cruz-Jentoft, A.J.; Sayer, A.A. Sarcopenia. *Lancet* **2019**, *393*, 2636–2646. [CrossRef]
2. Florence, C.S.; Bergen, G.; Atherly, A.; Burns, E.; Stevens, J.; Drake, C. Medical costs of fatal and nonfatal falls in older adults. *J. Am. Geriatr. Soc.* **2018**, *66*, 693–698. [CrossRef] [PubMed]
3. Brady, A.O.; Straight, C.R.; Evans, E.M. Body composition, muscle capacity, and physical function in older adults: An integrated conceptual model. *J. Aging Phys. Act.* **2014**, *22*, 441–452. [CrossRef] [PubMed]
4. Rodríguez-Sanz, D.; Tovaruela-Carrión, N.; López-López, D.; Palomo-López, P.; Romero-Morales, C.; Navarro-Flores, E.; Calvo-Lobo, C. Foot disorders in the elderly: A mini-review. *Dis. Mon.* **2018**, *64*, 64–91. [CrossRef] [PubMed]
5. Lukaski, H.C. Evolution of bioimpedance: A circuitous journey from estimation of physiological function to assessment of body composition and a return to clinical research. *Eur. J. Clin. Nutr.* **2013**, *67* (Suppl. 1), S2–S9. [CrossRef]
6. Lukaski, H.C.; Diaz, N.V.; Talluri, A.; Nescolarde, L. Classification of hydration in clinical conditions: Indirect and direct approaches using Bioimpedance. *Nutrients* **2019**, *11*, 809. [CrossRef]
7. Marini, E.; Campa, F.; Buffa, R.; Stagi, S.; Matias, C.N.; Toselli, S.; Sardinha, L.B.; Silva, A.M. Phase angle and bioelectrical impedance vector analysis in the evaluation of body composition in athletes. *Clin. Nutr.* **2020**, *39*, 447–454. [CrossRef]
8. Fukuda, D.H.; Stout, J.R.; Moon, J.R.; Smith-Ryan, A.E.; Kendall, K.L.; Hoffman, J.R. Effects of resistance training on classic and specific bioelectrical impedance vector analysis in elderly women. *Exp. Gerontol.* **2016**, *74*, 9–12. [CrossRef]
9. Campa, F.; Silva, A.M.; Toselli, S. Changes in phase angle and handgrip strength induced by suspension training in older women. *Int. J. Sports Med.* **2018**, *39*, 442–449. [CrossRef]
10. Souza, M.F.; Tomeleri, C.M.; Ribeiro, A.S.; Schoenfeld, B.J.; Silva, A.M.; Sardinha, L.B.; Cyrino, E.S. Effect of resistance training on phase angle in older women: A randomized controlled trial. *Scand. J. Med. Sci. Sports* **2016**, *27*, 1308–1316. [CrossRef]
11. Cunha, P.M.; Tomeleri, C.M.; Nascimento, M.A.; Nunes, J.P.; Antunes, M.; Nabuco, H.C.G.; Quadros, Y.; Cavalcante, E.F.; Mayhew, J.L.; Sardinha, L.B.; et al. Improvement of cellular health indicators and muscle quality in older women with different resistance training volumes. *J. Sports Sci.* **2018**, *36*, 2843–2848. [CrossRef] [PubMed]

12. Marini, E.; Buffa, R.; Saragat, B.; Coin, A.; Toffanello, E.D.; Berton, L.; Manzato, E.; Sergi, G. The potential of classic and specific bioelectrical impedance vector analysis for the assessment of sarcopenia and sarcopenic obesity. *Clin. Interv. Aging* **2012**, *7*, 585–591. [CrossRef] [PubMed]

13. Nunes, J.P.; Ribeiro, A.S.; Silva, A.M.; Schoenfeld, B.J.; dos Santos, L.; Cunha, P.M.; Nascimento, M.A.; Tomeleri, C.M.; Nabuco, H.C.G.; Antunes, M.; et al. Improvements in phase angle are related with muscle quality index after resistance training in older women. *J. Aging Phys. Act.* **2019**, *27*, 515–520. [CrossRef] [PubMed]

14. Norman, K.; Stobäus, N.; Pirlich, M.; Bosy-Westphal, A. Bioelectrical phase angle and impedance vector analysis—Clinical relevance and applicability of impedance parameters. *Clin. Nutr.* **2012**, *31*, 854–861. [CrossRef]

15. Beberashvili, I.; Azar, A.; Sinuani, I.; Shapiro, G.; Feldman, L.; Stav, K.; Sandbank, J.; Averbukh, Z. Bioimpedance phase angle predicts muscle function, quality of life and clinical outcome in maintenance hemodialysis patients. *Eur. J. Clin. Nutr.* **2014**, *68*, 683–689. [CrossRef]

16. Alves, F.D.; Souza, G.C.; Aliti, G.B.; Rabelo-Silva, E.R.; Clausell, N.; Biolo, A. Dynamic changes in bioelectrical impedance vector analysis and phase angle in acute decompensated heart failure. *Nutrition* **2015**, *31*, 84–89. [CrossRef]

17. Buffa, R.; Mereu, R.M.; Putzu, P.F.; Floris, G.; Marini, E. Bioelectrical impedance vector analysis detects low body cell mass and dehydration in patients with Alzheimer's disease. *J. Nutr. Health Aging* **2010**, *14*, 823–827. [CrossRef]

18. Campa, F.; Gatterer, H.; Lukaski, H.; Toselli, S. Stabilizing bioimpedance-vector-analysis measures with a 10-minute cold shower after running exercise to enable assessment of body hydration. *Int. J. Sports Physiol. Perform.* **2019**, *14*, 1006–1009. [CrossRef]

19. Campa, F.; Piras, A.; Raffi, M.; Trofè, A.; Perazzolo, M.; Mascherini, G.; Toselli, S. The effects of dehydration on metabolic and neuromuscular functionality during cycling. *Int. J. Environ. Res. Public Health* **2020**, *17*, 1161. [CrossRef]

20. Camina Martín, M.A.; de Mateo Silleras, B.; Nescolarde Selva, L.; Barrera Ortega, S.; Domínguez Rodríguez, L.; Redondo del Río, M.P. Bioimpedance vector analysis and conventional bioimpedance to assess body composition in older adults with dementia. *Nutrition* **2015**, *31*, 155–159. [CrossRef]

21. Ribeiro, A.S.; Nascimento, M.A.; Schoenfeld, B.J.; Nunes, J.P.; Aguiar, A.F.; Cavalcante, E.F.; Silva, A.M.; Fleck, S.J.; Sardinha, L.B.; Cyrino, E.S. Effects of single-set resistance training with different frequencies on cellular health indicator in older women. *J. Aging Phys. Act.* **2018**, *26*, 537–543. [CrossRef] [PubMed]

22. Ribeiro, A.S.; Schoenfeld, B.J.; dos Santos, L.; Nunes, J.P.; Tomeleri, C.M.; Cunha, P.M.; Sardinha, L.B.; Cyrino, E.S. Resistance training improves a cellular health parameter in obese older women: A randomized controlled trial. *J. Strength Cond. Res.* **2018**. [CrossRef]

23. Toselli, S.; Badicu, G.; Bragonzoni, L.; Spiga, F.; Mazzuca, P.; Campa, F. Comparison of the effect of different resistance training frequencies on phase angle and handgrip strength in obese women: A randomized controlled trial. *Int. J. Environ. Res. Public Health* **2020**, *17*, 1163. [CrossRef] [PubMed]

24. Ribeiro, A.S.; Schoenfeld, B.J.; Aguiar, A.F.; Nunes, J.P.; Cavalcante, E.F.; Cadore, E.L.; Cyrino, E.S. Effects of different resistance training systems on muscular strength and hypertrophy in resistance-trained older women. *J. Strength Cond. Res.* **2018**, *32*, 545–553. [CrossRef] [PubMed]

25. dos Santos, L.; Ribeiro, A.S.; Cavalcante, E.F.; Nabuco, H.C.G.; Antunes, M.; Schoenfeld, B.J.; Cyrino, E.S. Effects of modified pyramid system on muscular strength and hypertrophy in older women. *Int. J. Sports Med.* **2018**, *39*, 613–618. [CrossRef]

26. Basile, C.; Della-Morte, D.; Cacciatore, F.; Gargiulo, G.; Galizia, G.; Roselli, M.; Curcio, F.; Bonaduce, D.; Abete, P. Phase angle as bioelectrical marker to identify elderly patients at risk of sarcopenia. *Exp. Gerontol.* **2014**, *58*, 43–46. [CrossRef]

27. Sardinha, L.B.; Lohman, T.G.; Teixeira, P.; Guedes, D.P.; Going, S.B. Comparison of air displacment plethysmography with dual-energy X-ray absorptiometry and 3 field methods for estimating body composition in middle-aged men. *Am. J. Clin. Nutr.* **1998**, *68*, 786–793. [CrossRef]

28. Fragala, M.S.; Cadore, E.L.; Dorgo, S.; Izquierdo, M.; Kraemer, W.J.; Peterson, M.D.; Ryan, E.D. Resistance training for older adults: Position statement from the National Strength and Conditioning Association. *J. Strength Cond. Res.* **2019**, *33*, 2019–2052. [CrossRef]

29. Piccoli, A.; Pastori, G. *BIVA Software 2002*; Department of Medical and Surgical Sciences University of Padova: Padova, Italy, 2002.

30. Cohen, J. A power primer. *Psychol. Bull.* **1992**, *112*, 155–159. [CrossRef]

31. Schoenfeld, B.J. The mechanisms of muscle hypertrophy and their application to resistance training. *J. Strength Cond. Res.* **2010**, *24*, 2857–2872. [CrossRef]

32. Figueiredo, V.C.; de Salles, B.F.; Trajano, G.S. Volume for muscle hypertrophy and health outcomes: The most effective variable in resistance training. *Sports Med.* **2018**, *48*, 499–505. [CrossRef] [PubMed]

33. Schoenfeld, B.J.; Ogborn, D.; Krieger, J.W. Dose-response relationship between weekly resistance training volume and increases in muscle mass: A systematic review and meta-analysis. *J. Sports Sci.* **2017**, *35*, 1073–1082. [CrossRef] [PubMed]

34. Bosy-Westphal, A.; Danielzik, S.; Dörhöfer, R.-P.; Later, W.; Wiese, S.; Müller, M.J. Phase angle from bioelectrical impedance analysis: Population reference values by age, sex, and body mass index. *J. Parenter. Enter. Nutr.* **2006**, *30*, 309–316. [CrossRef] [PubMed]

35. Silva, A.M.; Matias, C.N.; Nunes, C.L.; Santos, D.A.; Marini, E.; Lukaski, H.C.; Sardinha, L.B. Lack of agreement of in vivo raw bioimpedance measurements obtained from two single and multi-frequency bioelectrical impedance devices. *Eur. J. Clin. Nutr.* **2019**, *73*, 1077–1083. [CrossRef]

36. Lacroix, A.; Hortobágyi, T.; Beurskens, R.; Granacher, U. Effects of supervised vs. unsupervised training programs on balance and muscle strength in older adults: A systematic review and meta-analysis. *Sports Med.* **2017**, *47*, 2341–2361. [CrossRef]

Article

Somatotype and Bioimpedance Vector Analysis: A New Target Zone for Male Athletes

Francesco Campa [1,*], Analiza M. Silva [2], Jacopo Talluri [3], Catarina N. Matias [2], Georgian Badicu [4] and Stefania Toselli [5]

[1] Department for Life Quality Studies, University of Bologna, 47921 Rimini, Italy
[2] Exercise and Health Laboratory, CIPER, Faculdade Motricidade Humana, Universidade de Lisboa, 1499-002 Cruz Quebrada, Portugal; analiza.monica@gmail.com (A.M.S.); cmatias@fmh.ulisboa.pt (C.N.M.)
[3] Department of Clinical Research and Development, Akern Ltd., 56121 Pisa, Italy; jacopo.talluri@akern.com
[4] Department of Physical Education and Special Motricity, University Transilvania of Brasov, 500068 Brasov, Romania; georgian.badicu@unitbv.ro
[5] Department of Biomedical and Neuromotor Sciences, University of Bologna, 40126 Bologna, Italy; stefania.toselli@unibo.it
* Correspondence: francesco.campa3@unibo.it

Received: 5 May 2020; Accepted: 25 May 2020; Published: 26 May 2020

Abstract: Background: Bioelectrical impedance vector analysis (BIVA) is a body composition assessment method based on the interpretation of the raw bioimpedance parameters. While it was initially proposed in clinical settings, its use in the sports field has grown considerably. The aim of this study was: (i) to explore the role of somatotype on BIVA patterns and (ii) to propose a new target zone to improve BIVA analysis in ball games athletes. **Methods:** One hundred and sixty-four male volleyball, soccer, and rugby players (age 26.2 ± 4.4 yrs; body mass index (BMI) 25.4 ± 2.4 kg/m^2) were included in this study. Somatotype and BIVA were measured from anthropometric and bioelectrical data, respectively. **Results:** Forty-six athletes were classified with an endomorphic mesomorphic somatotype, 26 showed a balanced mesomorphy, 55 were ectomorphic mesomorph, 10 resulted as mesomorph ectomorphs, 13 with a mesomorphic ectomorph somatotype, and in 14 athletes a balanced ectomorphy was assessed. The results of the Hotelling's T2 test showed significant differences in BIVA patterns for the endomorphic mesomorph group (p < 0.001) in comparison with all the other groups, while mesomorphic balanced athletes presented a more inclined vector compared to the athletes with a balanced ectomorphy (p < 0.003). In addition, the endomorphic mesomorph group showed a greater BMI (p < 0.001) with respect to the athletes grouped in the other somatotype categories. Discriminant analysis revealed two significant functions (p < 0.001). The first discriminant function primarily represented differences based on the bioelectrical standardized resistance parameter (R/H) measure, while the second function reflected differences based on the bioelectrical standardized reactance parameter (Xc/H). **Conclusions:** Athletes presenting a higher endomorphic component have a lower vector, whereas those with a larger mesomorphic component display higher vector inclinations on the R-Xc graph. We propose a new target zone to improve the interpretation of BIVA analysis in athletes engaged in team sports.

Keywords: anthropometry; BIVA; body composition; body shape; phase angle; vector length

---

1. Introduction

The assessment of body composition is essential for monitoring health status and the effects of exercise and nutritional regimen in athletes, as well as in the general population.

Morphologic features observed at the whole-body level of body composition analysis are relevant for sports performance [1,2]. In particular, useful information regarding body shape, proportionality, and composition can be obtained by assessing athletes' somatotype [3]. The anthropometric-based method proposed by Heath and Carter [4] is the most widely used for somatotyping individuals. The calculation is based on rating three numbers: the first number indicates the endomorphy, the second number the mesomorphy, and the third the ectomorphy component. The three calculated components are applied to a spherical triangle (a somatochart) on which the peaks reflect the predominance of one of the three components and the balance between the other two. In this regard, it should be considered that many athletes are not categorized as an exact stereotypical structure but have a preponderance that determines their position in the Heath-Carter proposed somatochart [3].

Nutrition and exercise have been demonstrated to influence somatotype [4]. Thus, anthropometry and body composition features play a crucial role in determining potential success in a particular sport [1,2,5]. Athletes vary in morphology and physical features according to the player's position in the team. Mesomorphic component is linked to individual sports that require muscle strength such as ball games and martial arts [6,7], while ectomorphy is predominant in runners, especially those involved in long distance [8]. Finally, athletes such as sumo wrestlers require a high amount of body fat, which determines a higher endomorphic component [9].

One of the most used methods to measure body composition in sports is the bioelectrical impedance analysis (BIA) [10]. The obtained raw impedance parameters allow two types of evaluations to be performed. The first is based on a quantitative assessment of body composition, where raw BIA measures (resistance (R) and reactance (Xc)) are inserted into prediction equations to then estimate the body composition variables, including fat mass (FM), fat free mass (FFM), total body water (TBW), and intracellular (ICW) and extracellular water (ECW). The second approach to BIA parameters is represented by a qualitative analysis where the raw impedance measurements are considered simultaneously through the bioimpedance vector analysis (BIVA). In BIVA, R and Xc are standardized for the height of the subject and plotted inside a graph as a point. The vector position can be compared with the percentiles of the reference population, or it is possible to study its displacements by collecting multiple measurements during a period of time (e.g., during a season, after exercise or an intervention program) [11–14]. Vector elongations or shortenings over time represent decreases or increases in TBW, while lateral displacements reflect changes in inclination, increases or decreases in the ICW/ECW ratio, and are therefore measured in the soft tissues [15,16].

BIVA was first proposed by Piccoli and colleagues in 1994 [17] and has become widely used in clinical settings. Based on the limited data collected, specific target zones were initially identified on the reference tolerance ellipses of the normal population [18–22]. In these ellipses, the athletes were positioned in the upper left side, while the obese population were placed in the lower left portion. Additionally, the right side of the vector analysis represented anorexics and cachexic who were positioned at the top and bottom, respectively. Although these target areas are still reported in recent clinical studies [23,24], latest evidence shows that heavier athletes can be placed in a portion of BIVA typically occupied by obese subjects [25–28]. Additionally, athletes show different BIVA patters than the normal population as well as among themselves in relation to the type of sport practiced. In particular, cyclists show a less inclined vector when compared to soccer players, indicating less muscle mass, while volleyball players show a shorter and inclined vector compared to these two groups due to a heavier body weight [26].

On this basis, body composition variables at the whole-body level (morphologic characteristics) could contribute to influencing the vector position on the R-Xc graph. A complete understanding of all contributing factors that change vector position would enable researchers and sports and nutrition professionals to correctly interpret BIVA, allowing for a more sustainable approach in assessing body composition. As mentioned above, this is due to the restrictive nature of the actual proposed target zone for athletes. In fact, those with a higher body weight appeared in the zone where obese subjects were categorized [23,24,26]. Therefore, a new target zone can help sports-related professionals and researchers when interpreting athletes' measurements in BIVA.

Kim et al. in 2010 [29] conducted a pilot study on 21 subjects including fashion models and dancers, demonstrating how BIVA patterns are influenced by somatotype. However, no study, so far, has investigated the influence of the somatotype on the bioimpedance vector in athletes varying in morphologic and physical features. Our hypothesis is that BIVA patterns are influenced by somatotype regardless of body composition characteristics at the molecular level in athletes. Therefore, the aim of this study was to explore the role of somatotype on BIVA patterns and to propose a new target zone in order to improve BIVA analysis in male ball game athletes.

2. Materials and Methods

2.1. Subjects

This was a cross-sectional observational study on 164 athletes (88 volleyball, 41 soccer, and 35 rugby players) belonging to 7 professional Italian teams participating in Series A2, Series B, and Series A divisions in volleyball, soccer, and rugby, respectively (age 26.2 ± 4.4 yrs; body mass index (BMI) 25.4 ± 2.4 kg/m^2). The following inclusion criteria were used: (1) a minimum of 10 h to a maximum of 13 h of training per week; (2) tested negative for performance-enhancing drugs; and (3) not taking any medications. All participants gave informed consent after receiving a detailed description of the study procedures. Athletes were tested in the morning (9.00 AM) during the off-season period at the facilities of each team. The project was conducted according to the Declaration of Helsinki and was authorized by the Bioethics Committee of the University of Bologna.

2.2. Procedures

The subjects came to the sport center refraining from vigorous exercise at least 15 h prior, no caffeine and alcohol intake during the preceding 24 h, and consuming a normal evening meal the night before. All athletes were tested to ensure a well-hydrated state using the urine specific gravity test (refractometer Urisys 1100; Roche Diagnostics), from a fasting baseline urine sample, according to Armstrong et al. [30]. A urine specific gravity value <1.022 for the first urine was used to categorize euhydration.

The anthropometric traits were body mass, height, humerus and femur breadths, upper arm (relaxed and contracted), calf, and thigh girths. All anthropometric measurements were taken according to standard methods [31]. Height was recorded to the nearest 0.1 cm using a stadiometer (Raven Equipment Ltd., Great Donmow, UK) and body mass was measured to the nearest 0.1 kg using a high-precision mechanical scale (Seca, Basel, Switzerland). BMI was calculated as the ratio of body weight to height squared (kg/m^2). Girths were taken to the nearest 0.1 cm using a tape measure (GMP, Zürich, Switzerland). Breadths were measured to the nearest 0.1 cm using a sliding caliper (GMP, Zürich, Switzerland). Skinfold thicknesses at 8 sites (biceps, triceps, subscapular, supraspinal, suprailiac, lateral calf, medial calf, and thigh) were measured to the nearest 0.1 mm using a Lange skinfold caliper (Beta technology Inc., Cambridge, Maryland). The muscle area of the thigh (TTM), calf (CMA) and upper arm (UMA), as well as the fat area of the thigh, calf, and upper arm (UFA) were calculated according to Frisancho [32].

The impedance measurements were performed by a phase-sensitive single-frequency bioimpedance analyzer (BIA 101 Anniversary, Akern, Florence, Italy), which applies an alternating current of 400 microamperes at 50 kHz. The subjects were in the supine position with a leg opening of 45° compared to the median line of the body and the upper limbs positioned 30° away from the trunk. After cleansing the skin with alcohol, two Ag/AgCl low-impedance electrodes (Biatrodes, Akern Srl, Florence, Italy) were placed on the back of the right hand and two electrodes on the corresponding foot, with a distance of 5 cm between each other. Vector length (Z) was calculated as (adjusted R^2 + adjusted Xc^2)$^{0.5}$ and phase angle (PhA) as the arctangent of Xc/R x 180/. BIVA was carried out using the classic methods, e.g., normalizing Z, R, and Xc for height in meters [33].

The BIA parameters obtained for each subject were used to calculate the FM, FFM, TBW, ICW, and ECW values using the software Bodygram®. Somatotype components were calculated, and athletes grouped according to the Heath-Carter method [34].

2.3. Statistical Analysis

The mean standard deviation was calculated for each variable. To verify the normality of the data, the Shapiro–Wilk test was applied. Univariate analysis of variance for multiple comparisons was performed. When a significant F ratio was obtained, the Bonferroni post hoc test was used to assess the differences among the groups ($p < 0.003$). The two-sample Hotelling's T^2 test was used to compare the mean impedance vectors of the different somatotype categories. Discriminant function analysis (stepwise criteria) was then applied to R/H, Xc/H, Z/H, and PhA to classify athletes into the different somatotype categories according to bioelectric features. Data was analyzed with IBM SPSS Statistics (version 24.0; IBM, Chicago, IL).

3. Results

The descriptive statistics of anthropometric, body composition, bioelectrical, and somatotype components are presented in Table 1.

Table 1. Anthropometric, body composition, bioelectric, and somatotype data of the athletes.

Variable	Soccer (n = 41)	Volleyball (n = 88)	Rugby (n = 35)
Age (years)	26.3 ± 3.2	26.5 ± 5.6	25.9 ± 4.3
Height (m)	180.6 ± 7.0	194.1 ± 10.1	184.2 ± 8.4
Weight (kg)	74.7 ± 8.8	89.5 ± 10.4	100.7 ± 15.7
BMI (kg/m^2)	22.8 ± 1.7	23.8 ± 1.9	29.6 ± 3.5
FM (%)	11.8 ± 2.6	13.5 ± 2.1	15.5 ± 4.3
FM (kg)	8.9 ± 2.6	12.2 ± 2.8	16.1 ± 6.3
FFM (kg)	65.7 ± 7.0	77.3 ± 8.6	84.6 ± 10.6
TBW (l)	54.1 ± 3.4	59.8 ± 4.0	64.2 ± 6.1
ECW (l)	12.1 ± 1.1	14.1 ± 1.3	15.3 ± 2.1
ICW (l)	42.0 ± 2.3	45.7 ± 2.7	48.9 ± 4.1
UMA (cm^2)	59.1 ± 10.3	70.8 ± 14.7	83.7 ± 14.1
UFA (cm^2)	8.6 ± 2.4	9.9 ± 3.3	12.9 ± 3.7
CMA (cm^2)	104.7 ± 16.7	104.9 ± 14.9	122.1 ± 40.1
CFA (cm^2)	9.8 ± 2.3	12.1 ± 4.3	14.7 ± 5.5
TMA (cm^2)	200.2 ± 29.2	222.8 ± 37.5	260.1 ± 57.1
TFA (cm^2)	22.7 ± 8.2	28.9 ± 13.4	32.3 ± 11.0
R/H (Ohm/m)	255.6 ± 21.3	236.4 ± 23.4	213.6 ± 21.8
Xc/H (Ohm/m)	35.3 ± 3.8	31.8 ± 3.9	30.2 ± 4.2
Z/H (Ohm/m)	258.1 ± 21.5	238.5 ± 24.7	215.7 ± 22.1
PhA (°)	7.9 ± 0.5	7.7 ± 0.6	8.0 ± 0.8
Endomorphy	1.6 ± 0.3	2.0 ± 0.7	2.1 ± 0.7
Mesomorphy	4.7 ± 0.9	4.0 ± 1.3	6.0 ± 1.1
Ectomorphy	2.9 ± 0.8	3.2 ± 1.1	0.9 ± 0.3

Note: Data are presented as mean ± SD, BMI = body mass index, FM = fat mass, FFM = fat free mass, TBW = total body water, ECW = extracellular water, ICW = intracellular water, UMA = upper arm muscle area, UFA = upper arm fat area, CMA = calf muscle area, CFA = calf fat area, TMA = thigh muscle area, TFA = thigh fat area, R/H = resistance standardized for height, Xc/H = reactance standardized for height, Z/H = vector length standardized for height, PhA = phase angle.

Six types of somatotypes have been recognized based on the athlete's position on the somatochart (Figure 1):

1. Endomorphic mesomorph: mesomorphy was dominant and endomorphy was greater than ectomorphy (more than 0.5 units).

2. Balanced mesomorph: mesomorphy was dominant, and endomorphy and ectomorphy were similar (no difference or <0.5 units).
3. Ectomorphic mesomorph: mesomorphy was dominant and ectomorphy was greater than endomorphy (more than 0.5 units).
4. Mesomorph ectomorph: endomorphy and ectomorphy were similar (no difference or <0.5 units).
5. Mesomorphic ectomorph: ectomorphy was dominant and mesomorphy was greater than endomorphy (more than 0.5 units).
6. Balanced ectomorphy: ectomorphy was dominant, and endomorphy and ectomorphy were similar (no difference or <0.5 units).

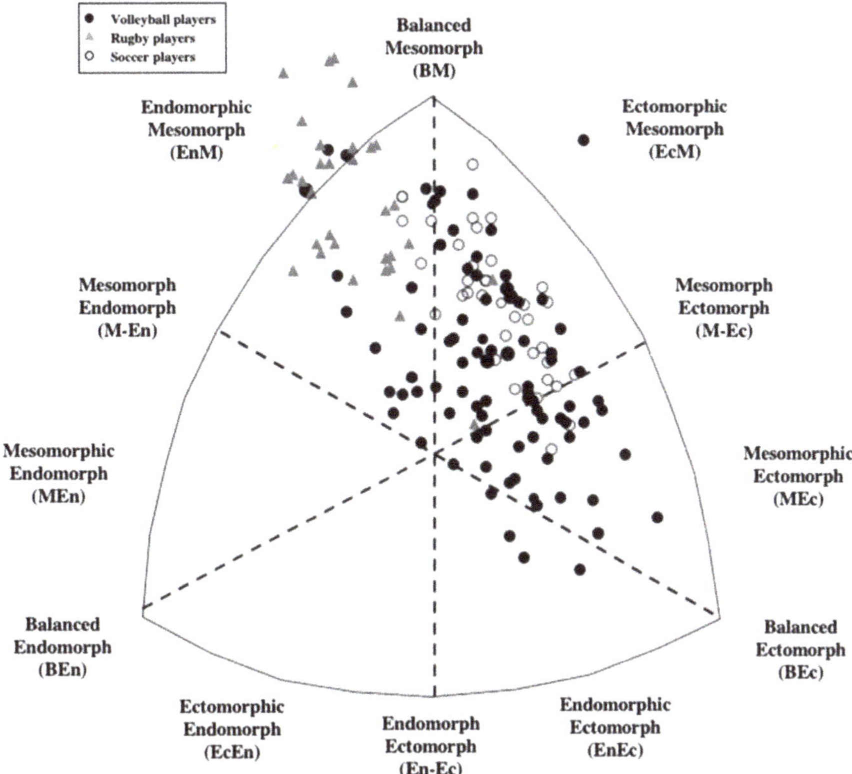

Figure 1. Representation of the athletes' somatotype.

Endomorphic mesomorph athletes included 33 rugby, 10 volleyball, and 3 soccer players; balanced mesomorph athletes included 19 volleyball and 7 soccer players; ectomorphic mesomorph athletes included 22 volleyball, 31 soccer, and 2 rugby players; mesomorph ectomorph athletes included 10 volleyball players; mesomorphic ectomorph athletes included 13 volleyball players; and balanced ectomorph athletes included 14 volleyball players. Overall, rugby players showed on average an endomorphic mesomorph somatotype, while soccer and volleyball player presented an ectomorphic mesomorph profile.

Table 2 provides the comparison of anthropometric, body composition, bioelectrical, and somatotype features. Endomorphic mesomorph athletes showed a higher BMI than all the other groups ($p < 0.003$), and additionally presented greater endomorphy, mesomorphy, and ectomorphy components compared with the athletes of different somatotype.

Table 2. Anthropometric, body composition, and bioelectrical data of the athletes separated by somatotype.

Variable	Endomorphic Mesomorph (n = 46)	Balanced Mesomorph (n = 26)	Ectomorphic Mesomorph (n = 55)	Mesomorph Ectomorph (n = 10)	Mesomorphic Ectomorph (n = 13)	Balanced Ectomorph (n = 14)	ANOVA p	
Height (m)	183.6 ± 8.7 [4,5,6]	186.7 ± 12.3 [5]	186.7 ± 9.4 [4,5]	199.9 ± 3.4 [1,3]	201.1 ± 5.3 [1,2,3]	195.9 ± 9.4 [1]	12.7	<0.01
Weight (kg)	96.6 ± 15.6 [3]	85.0 ± 14.2	81.2 ± 13.1 [1]	92.2 ± 6.4	90.9 ± 5.8	88.1 ± 10.6	7.3	<0.01
BMI (kg/m^2)	28.6 ± 3.7 [2,3,4,5,6]	24.2 ± 1.8 [1]	23.1 ± 1.7 [1]	23 ± 1 [1]	22.4 ± 1.1 [1]	22.9 ± 1.6 [1]	32.8	<0.01
FM (%)	15.1 ± 3.9 [3]	13.2 ± 1.8	11.8 ± 2.2 [1]	14.2 ± 2.4	14.2 ± 1.9	13.9 ± 2.6	7.1	<0.01
FM (kg)	15 ± 5.9 [2]	11.3 ± 2.9	9.7 ± 2.9 [1]	13.1 ± 3	12.9 ± 1.8	12.4 ± 3.2	9.1	<0.01
FFM (kg)	81.5 ± 10.8 [3]	73.6 ± 11.6	71.4 ± 10.8 [1]	79 ± 4.4	78 ± 5.5	75.7 ± 8.5	5.4	<0.01
TBW (l)	62.5 ± 6 [3]	58.1 ± 5.4	56.6 ± 5 [1]	60.8 ± 2.4	60.3 ± 2.2	59.3 ± 4	7.3	<0.01
ECW (l)	14.7 ± 1.9 [3]	13.3 ± 1.8	12.8 ± 1.6 [1]	14.2 ± 0.81	14.1 ± 0.74	13.7 ± 1.3	7.3	<0.01
ICW (l)	47.8 ± 4.0 [3]	44.8 ± 3.6	43.8 ± 3.3 [1]	46.6 ± 1.6	46.2 ± 1.5	45.6 ± 2.7	7.3	<0.01
UMA (cm^2)	80.0 ± 14.4 [3]	68.2 ± 12.5	66.0 ± 14.7 [1]	67.1 ± 8.4	70.9 ± 12.8	64.1 ± 10.5	6.5	<0.01
UFA (cm^2)	12.5 ± 3.5 [3,5]	11.1 ± 3.1 [3,]	8.1 ± 2.3 [1,2]	10.1 ± 4.4	8.6 ± 2.4 [1]	10.8 ± 3.3	11.1	<0.01
CMA (cm^2)	117.7 ± 36.1	103.3 ± 13.6	108.8 ± 17.4	101.0 ± 16.0	99.5 ± 13.6	99.9 ± 12.5	2.6	0.02
CFA (cm^2)	14.1 ± 5.7 [3]	12.5 ± 3.9	10.1 ± 2.9 [1]	11.6 ± 3.6	11.6 ± 3.7	12.9 ± 3.9	4.7	<0.01
TMA (cm^2)	255.9 ± 61.4 [3,6]	220.65 ± 25.1	212.8 ± 33.7 [1]	212.3 ± 16.7	210.0 ± 26.8	203.3 ± 32.2 [1]	7.3	<0.01
TFA (cm^2)	34.9 ± 16.0 [3]	31.1 ± 12.7	22.7 ± 8.7 [1]	29.8 ± 10.2	23.4 ± 10.9	26.5 ± 9.2	4.9	<0.01
R/H (Ohm/m)	220.0 ± 23.1 [3]	240.4 ± 26.9	242.7 ± 27.1 [1]	239.3 ± 15.5	245.0 ± 21.4	246.7 ± 25.0	5.6	<0.01
Xc/H (Ohm/m)	31.3 ± 4.4	32.8 ± 4.3	33.1 ± 4.7	31.6 ± 2.3	32.3 ± 2.9	31.7 ± 3.7	1.1	0.39
Z/H (Ohm/m)	222.3 ± 23.3 [3]	242.7 ± 27.1	245.1 ± 27.4 [1]	241.5 ± 15.6	247.2 ± 21.6	248.8 ± 25.3	5.5	<0.01
PhA (°)	8.1 ± 0.74 [6]	7.7 ± 0.65	7.7 ± 0.58	7.5 ± 0.43	7.5 ± 0.40	7.3 ± 0.39 [1]	4.8	<0.01
Endomorphy	2.8 ± 0.59 [2,3,4,5,6]	2.3 ± 0.50 [1,3,5]	1.5 ± 0.28 [1,2,6]	1.8 ± 0.64 [1]	1.4 ± 0.41 [1,2,6]	2.1 ± 0.44 [1,3,5]	43.8	<0.01
Mesomorphy	5.9 ± 1.1 [2,3,4,5,6]	4.5 ± 1.1 [1,5,6]	4.6 ± 0.82 [1,5,6]	3.6 ± 0.43 [1]	3.1 ± 0.39 [1,2,3]	2.2 ± 0.54 [1,2,3]	47.7	<0.01
Ectomorphy	0.9 ± 0.3 [2,3,4,5,6]	2.3 ± 0.51 [1,3,4,5,6]	3.1 ± 0.57 [1,2,5,6]	3.6 ± 0.27 [1,2]	4.4 ± 0.67 [1,2,3]	4.2 ± 0.56 [1,2,3]	132.0	<0.01

Note: Data are presented as mean ± SD, BMI = body mass index, FM = fat mass, FFM = fat free mass, TBW = total body water, ECW = extracellular water, ICW = intracellular water, UMA = upper arm muscle area, UFA = upper arm fat area, CMA = calf muscle area, CFA = calf fat area, TMA = thigh muscle area, TFA = thigh fat area, R/H = resistance standardized for height, Xc/H = reactance standardized for height, Z/H = vector length standardized for height, PhA = phase angle.

1. Differences ($p < 0.003$) compared with the Endomorphic Mesomorph group;
2. Differences ($p < 0.003$) compared with the Balanced Mesomorph group;
3. Differences ($p < 0.003$) compared with the Ectomorphic Mesomorph group;
4. Differences ($p < 0.003$) compared with the Mesomorph Ectomorph group;
5. Differences ($p < 0.003$) compared with the Mesomorphic Ectomorph group;
6. Differences ($p < 0.003$) compared with the Balanced Ectomorph group.

Figure 2 shows the vector displacement of the six somatotype groups on the R-Xc graph. Endomorphic mesomorph athletes showed a shorter vector in comparison with the other groups ($p < 0.001$), while athletes with a balanced mesomorphy presented a more inclined vector compared to athletes with a balanced ectomorphy ($p < 0.001$) (Figure 3). No statistically significant differences were found between the other groups. Figure 3 shows the results of the Hotelling's T^2 test; separate 95% confidence ellipses indicated a significant vector difference.

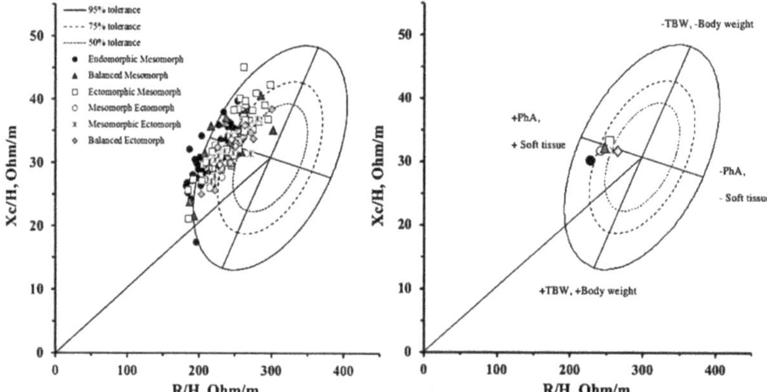

Figure 2. Scatter plots of the individual (**on the left side**) and mean (**on the right side**) impedance vectors, divided by somatotype categories and plotted on the tolerance ellipses of the general population [18].

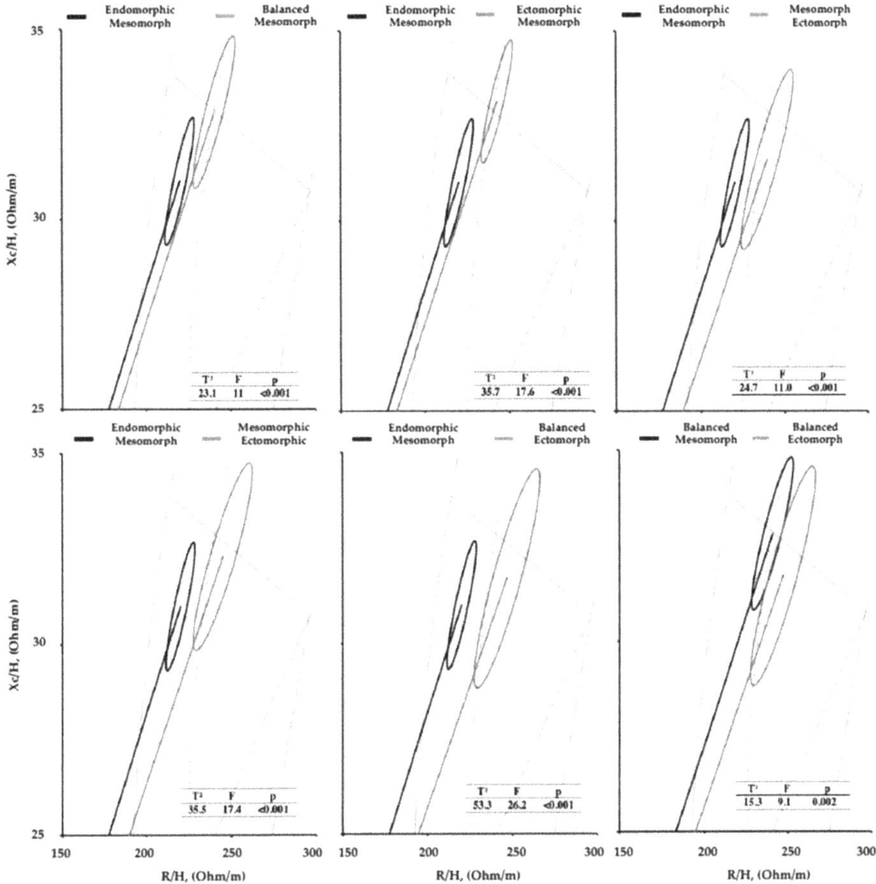

Figure 3. Mean impedance vectors with the 95% confidence ellipses for somatotype groups that showed significant differences ($p < 0.003$). The Hotelling's T^2 test results are included.

Discriminant analysis revealed two significant functions ($p < 0.001$) (Table 3). The first discriminant function primarily represented differences based on the R/H. The second discriminant function represented differences based on the Xc/H proportion. All of the variance explained by the model is due to the first two discriminant functions. Based on values of Wilk's lambda, the first discriminant function accounted for 93.1% (eigenvalue = 0.326) of the total variance, while the second discriminant function explained 6.9% (eigenvalue = 0.024) of the remaining variance. Figure 4 represents group centroid distances between somatotype categories for both discriminant functions.

Table 3. Results of stepwise discriminant analyses.

Step	Entred	Wilks' Lambda	F	p
1	R/H (Ohm/m)	0.849	5.63	<0.001
2	Xc/H (Ohm/m)	0.736	5.19	<0.001

Note: R/H = resistance standardized for height, Xc/H = reactance standardized for height.

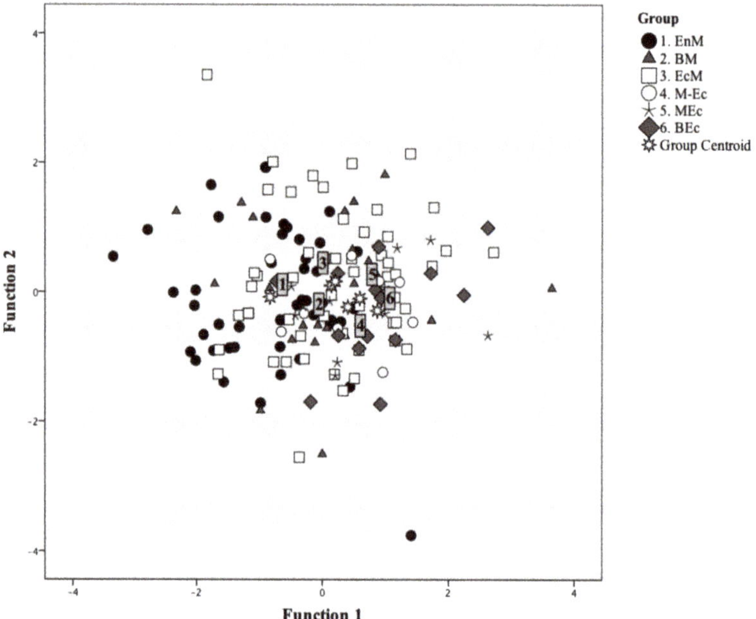

Figure 4. Canonical discriminant functions. EnM = Endomorphic Mesomorph, BM = Balanced Mesomorph, EcM = Ectomorphic Mesomorph, M-Ec = Mesomorph Ectomorph, MEc = Mesomorphic Ectomorph, BEc = Balanced Ectomorph.

4. Discussion

This study showed the role of somatotype on BIVA patterns in athletes. In particular, when the endomorphic component was higher, the vector resulted in a lower position on the R-Xc graph, while a larger mesomorphic component resulted in a higher vector inclination and, thus, an increased PhA. We also proposed a new target zone that improves BIVA analysis in this particular sample of male athletes. This new target zone extends the area where athletes can be positioned, showing how the three somatotype components (I = endomorphy, II = mesomorphy, and III = ectomorphy) contribute to the vector placements (Figure 5).

The athletes tested in this study were involved in one of three typical team sports where optimal body composition features are necessary [1,2]. However, some of these athletes had a greater BMI and an FM% slightly higher than average compared to their teammates. These resulted in an athlete with a somatotype where the mesomorphic component was predominant, but the endomorphic value was greater than the ectomorphic measurement. In fact, 94% of the rugby players evaluated showed an endomorphic mesomorph somatotype. This can be an advantage for rugby players, especially those who play the position of pylon, where a greater weight is favorable for their performance [35,36]. This group of athletes presented a greater BMI but also tended to have a greater FM distribution in the arm, thigh, and calf segments. This trend reflected a significantly shorter vector in comparison to the other groups. The analysis of the average impedance vectors also showed a significant difference between the groups with balanced mesomorphy and ectomorphy. In particular, subjects with a greater mesomorphic component showed a vector significantly more inclined and a greater PhA when compared to athletes with a balanced ectomorphic somatotype. Although not statistically significant, athletes with a balanced mesomorphy showed a tendency for a greater TMA, CMA, and UMA as an expected result of a higher skeletal muscle component. Vector length has recently been associated with the amount of intra- and extracellular fluids obtained through dilution techniques as the reference method, while the inclination that determines the PhA is directly associated with the ICW/ECW ratio [15,16].

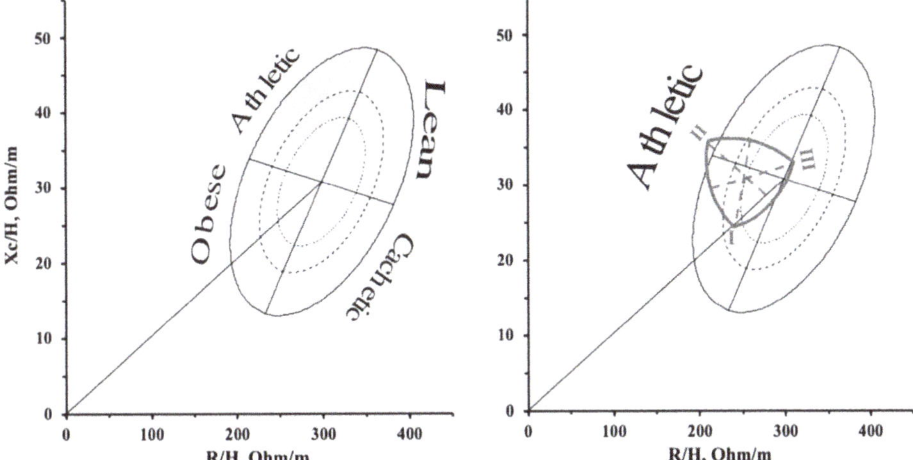

Figure 5. (**Left side**) The target zone on the R-Xc graph proposed earlier for the general population [18–22]. (**Right side**) The new target zone proposed for athletes.

The R/H variable stands for the first discriminant function that primarily represents differences between athletes of different somatotypes. The main difference in BIVA patterns among the athletes examined in this study was linked to the different morphology of the athletes categorized with an endomorphic mesomorph somatotype. This was reflected in a lower and more inclined vector in the R-Xc graph observed in these athletes compared to the other somatotype classes. This is in accordance with previous studies that the major differences among high-level and elite athletes can be found considering R/H and X/C simultaneously in BIVA [26,27,37], which is due to the inverse association of R with TBW, while Xc is directly proportionate to the cellular density [17]. Therefore, while ICW/ECW ratio can be similar among athletes, the absolute values of ICW and ECW can differ, reflecting the different body structures. In this regard, BIVA has a higher efficiency when evaluating body composition than only the interpretation on PhA [26]. This is due to BIVA taking into consideration the vector inclination, which reflects PhA and, therefore, ICW/ECW ratio, in addition to the vector length, which represents TBW [15]. In this study, the endomorphic mesomorph athletes showed that both values for R/H and Xc/H tended to be lower. These values, when analyzed simultaneously through the Hotelling's T^2 test, resulted in a significantly shorter and inclined vector in the R/Xc graph in comparison to the other athletes.

The athletes involved in this study showed a somatotype in line with earlier studies conducted in volleyball, rugby, and soccer players [2,38,39]. In particular, rugby players were found to have an endomorphic mesomorph somatotype, while soccer players and volleyball players presented an ectomorphic mesomorph somatotype. Some athletes showed different morphological characteristics compared to those of their group average, but this is due to the different requirements of the positions played, as already recognized [39,40]. In line with reports by previous studies [26,27] athletes were positioned outside the 50th percentile of the tolerance ellipses, indicating greater ICW/ECW ratio and soft tissues compared to the general population. On the other hand, for the same type of sport, differences in the inclination angle between the impedance vector and the standardized resistance were found according to competitive levels, with greater PhA observed in top-level athletes [26,27,41].

Considering that, in our study, athletes underwent a single measurement, the vector position in the R-Xc graph reflects body composition features. On the contrary, studies in which more than one

BIVA test has been collected have shown that the vector may undergo temporary changes. Exercise that causes fluid loss, for example, generates a vector lengthening in the short term [42,43]. On the other hand, Pollastri et al. [44] demonstrated that during a cycling tour, vector shortening is accompanied by an increase in body weight. This condition could have identified water retention and edema, a scenario that also occurs following muscle injuries. In this regard, Nescolarde and colleagues [45] conducted an observational prospective study with professional soccer players that presented muscle injuries. The authors monitored bioimpedance parameters using a localized approach and observed decreases in R as a result of edema in the injured limb, and decreases in PhA as a consequence of increases in the ECW/ICW ratio [45]. Therefore, if the somatotype does not change over time, the vector position can be influenced by the loss or gain of body fluids that can occur in athletes as a result of a variety of different conditions.

BIVA patterns can also change during the competitive season, which highlights this approach as a useful tool in providing information on training adaptations occurring throughout the season [46]. Shortening and lengthening in the vector inclination are instead obtained with the increase of somatic maturation, reflecting changes in body composition that occur alongside growth [47]. Otherwise, in response to following an exercise program, there is a lengthening and a shift to the left of the vector due to a reduction in extracellular fluids and an increase in ICW/ECW ratio. Finally, an anomalous case of vector alteration can be represented by measurements taken immediately after exercise, where a high body temperature can lead to flaws in BIA [48].

Some limitations in this study should be considered. First, our results are applicable to single-frequency BIA equipment. In fact, different results in measuring raw BIA parameters are obtained using devices that work on single- or multi-frequency [49]. In addition, our results are only generalized for male athletes involved in team sports. Further studies should be conducted in other sports, with a different body composition profile, and also in the female athletic population. Lastly, BIVA should not be considered as an alternative tool for measuring the somatotype, but as an additional approach to evaluating athletes considering that morphology, together with other body composition parameters, influences the vector position in the R-Xc graph.

5. Conclusions

This study demonstrates how the different components of the somatotype are associated with the BIVA patterns. A greater endomorphic component is associated with a shorter vector, while a larger mesomorphic component is associated with an increase in the vector inclination in the R-Xc graph. The vector position is therefore determined by the body composition characteristics at the molecular and whole-body levels. A new target zone should be considered to improve the interpretation of BIVA analysis in a male athletic population involved in team sports.

Author Contributions: Conceptualization, F.C. and S.T.; methodology, F.C., A.M.S., C.N.M., J.T., and S.T.; software, F.C. and S.T.; formal analysis, F.C. and S.T.; investigation, F.C. and S.T.; data curation, A.M.S., C.N.M., F.C., J.T., G.B., and S.T.; writing—original draft preparation, F.C., A.M.S., C.N.M., J.T., G.B., and S.T.; visualization, F.C., G.B., J.T., and S.T.; supervision, S.T., J.T.; project administration, F.C. All authors have read and agreed to the published version of the manuscript.

Funding: This research received no external funding.

Acknowledgments: The authors are grateful to all the athletes who took part in this study.

Conflicts of Interest: The authors declare no conflict of interest.

References

1. Campa, F.; Semprini, G.; Júdice, P.B.; Messina, G.; Toselli, S. Anthropometry, physical and movement features, and repeated-sprint ability in soccer players. *Int. J. Sports Med.* **2019**, *40*, 100–109. [CrossRef]
2. Toselli, S.; Campa, F. Anthropometry and functional movement patterns in elite male volleyball players of diferent competitive levels. *J. Strength Cond. Res.* **2018**, *32*, 2601–2611. [CrossRef] [PubMed]

3. Carter, J.E.L.; Heath, B.H. *Somatotyping Development and Applications*; Cambridge University Press: Cambridge, UK, 1990.
4. Carter, J.E.; Phillips, W.H. Structural changes in exercising middle-aged males during a 2-year period. *J. Appl. Physiol.* **1969**, *27*, 787–794. [CrossRef] [PubMed]
5. Choudhary, S.; Singh, S.; Singh, I.; Varte, L.R.; Sahani, R.; Rawat, S. Somatotypes of Indian Athletes of Different Sports. *Online J Health Allied Sci.* **2019**, *18*, 6.
6. Gualdi, R.E.; Graziani, I. Anthropometric somatotype of Italian sport participants. *J. Sports Med. Phys. Fit.* **1993**, *33*, 282–291. [PubMed]
7. Raković, A.; Savanović, V.; Stanković, D.; Pavlović, R.; Simeonov, A.; Petković, E. Analysis of the elite athletes' somatotypes. *Acta Kinesiol.* **2015**, *9*, 47–53.
8. Sánchez Muñoz, C.; Muros, J.J.; López Belmonte, Ó.; Zabala, M. Anthropometric characteristics, body composition and somatotype of elite male young runners. *Int. J. Environ. Res. Public Health* **2020**, *17*, 674. [CrossRef]
9. Sánchez-Muñoz, C.; Zabala, M.; Williams, K. Anthropometric variables and its usage to characterise elite youth athletes. In *Handbook of Anthropometry*; Preedy, V., Ed.; Springer: New York, NY, USA, 2012; ISBN 978-1-4419-1787-4.
10. Castizo-Olier, J.; Irurtia, A.; Jemni, M.; Carrasco-Marginet, M.; Fernández-García, R.; Rodríguez, F.A. Bioelectrical impedance vector analysis (BIVA) in sport and exercise: Systematic review and future perspectives. *PLoS ONE* **2018**, *13*, e0197957. [CrossRef]
11. Campa, F.; Matias, C.N.; Marini, E.; Heymsfield, S.B.; Toselli, S.; Sardinha, L.B.; Silva, A.M. Identifying athlete body-fluid changes during a competitive season with bioelectrical impedance vector analysis. *Int. J. Sports Physiol. Perform.* **2020**, *15*, 361–367. [CrossRef]
12. Campa, F.; Silva, A.M.; Toselli, S. Changes in phase angle and handgrip strength induced by suspension training in older women. *Int. J. Sports Med.* **2018**, *39*, 442–449. [CrossRef]
13. Souza, M.F.; Tomeleri, C.M.; Ribeiro, A.S.; Schoenfeld, B.J.; Silva, A.M.; Sardinha, L.B.; Cyrino, E.S. Effect of resistance training on phase angle in older women: A randomized controlled trial. *Scand. J. Med. Sci. Sports* **2017**, *27*, 1308–1316. [CrossRef] [PubMed]
14. Toselli, S.; Badicu, G.; Bragonzoni, L.; Spiga, F.; Mazzuca, P.; Campa, F. Comparison of the effect of different resistance training frequencies on phase angle and handgrip strength in obese women: A randomized controlled trial. *Int. J. Environ. Res. Public Health* **2020**, *17*, 1163. [CrossRef] [PubMed]
15. Marini, E.; Campa, F.; Buffa, R.; Stagi, S.; Matias, C.N.; Toselli, S.; Sardinha, L.B.; Silva, A.M. Phase angle and bioelectrical impedance vector analysis in the evaluation of body composition in athletes. *Clin. Nutr.* **2020**, *39*, 447–454. [CrossRef]
16. Francisco, R.; Matias, C.N.; Santos, D.A.; Campa, F.; Minderico, C.S.; Rocha, P.; Heymsfield, S.B.; Lukaski, H.; Sardinha, L.B.; Silva, A.M. The predictive role of raw bioelectrical impedance parameters in water compartments and fluid distribution assessed by dilution techniques in athletes. *Int. J. Environ. Res. Public Health* **2020**, *17*, 759. [CrossRef] [PubMed]
17. Piccoli, A.; Rossi, B.; Pillon, L.; Bucciante, G. A new method for monitoring body fluid variation by bioimpedance analysis: The RXc graph. *Kidney Int.* **1994**, *46*, 534–539. [CrossRef] [PubMed]
18. Piccoli, A.; Nigrelli, S.; Caberlotto, A.; Bottazzo, S.; Rossi, B.; Pillon, L.; Maggiore, Q. Bivariate normal values of the bioelectrical impedance vector in adult and elderly populations. *Am. J. Clin. Nutr.* **1995**, *61*, 269–270. [CrossRef] [PubMed]
19. Piccoli, A.; Pillon, L.; Dumler, F. Impedance vector distribution by sex, race, body mass index, and age in the United States: Standard reference intervals as bivariate Z scores. *Nutrition* **2002**, *18*, 153–167. [CrossRef]
20. Rösler, A.; Lehmann, F.; Krause, T.; Wirth, R.; Von Renteln-Kruse, W. Nutritional and hydration status in elderly subjects: Clinical rating versus bioimpedance analysis. *Arch. Gerontol. Geriatr.* **2010**, *50*, e81–e85. [CrossRef]
21. Bronhara, B.; Piccoli, A.; Pereira, J.C.R. Fuzzy linguistic model for bioelectrical impedance vector analysis. *Clin. Nutr.* **2012**, *31*, 710–716. [CrossRef]
22. Maioli, M.; Toso, A.; Leoncini, M.; Musilli, N.; Bellandi, F.; Rosner, M.H.; McCullough, P.A.; Ronco, C. Pre-procedural bioimpedance vectorial analysis of fluid status and prediction of contrast-induced acute kidney injury. *J. Am. Coll. Cardiol.* **2014**, *63*, 1387–1394. [CrossRef]

23. González-Correa, C.H. Body composition by bioelectrical impedance analysis. In *Bioimpedance in Biomedical Applications and Research*; Simini, F., Bertemes-Filho, P., Eds.; Springer: Cham, Switzerland, 2018; ISBN 978-3-319-74387-5.

24. Nwosu, A.C.; Mayland, C.R.; Mason, S.; Cox, T.F.; Varro, A.; Stanley, S.; Ellershaw, J. Bioelectrical impedance vector analysis (BIVA) as a method to compare body composition differences according to cancer stage and type. *Clin. Nutr.* **2019**, *30*, 59–66. [CrossRef] [PubMed]

25. Campa, F.; Toselli, S. Bioimpedance vector analysis of elite, subelite, and low-level male volleyball players. *Int. J. Sports Physiol. Perform.* **2018**, *13*, 1250–1253. [CrossRef] [PubMed]

26. Campa, F.; Matias, C.; Gatterer, H.; Toselli, S.; Koury, J.C.; Andreoli, A.; Melchiorri, G.; Sardinha, L.B.; Silva, A.M. Classic bioelectrical impedance vector reference values for assessing body composition in male and female athletes. *Int. J. Environ. Res. Public Health* **2019**, *16*, 5066. [CrossRef] [PubMed]

27. Micheli, M.L.; Pagani, L.; Marella, M.; Gulisano, M.; Piccoli, A.; Angelini, F.; Burtscher, M.; Gatterer, H. Bioimpedance and impedance vector patterns as predictors of league level in male soccer players. *Int. J. Sports Physiol. Perform.* **2014**, *9*, 532–539. [CrossRef] [PubMed]

28. Carrasco-Marginet, M.; Castizo-Olier, J.; Rodríguez-Zamora, L.; Iglesias, X.; Rodríguez, F.A.; Chaverri, D.; Brotons, D.; Irurtia, A. Bioelectrical impedance vector analysis (BIVA) for measuring the hydration status in young elite synchronized swimmers. *PLoS ONE* **2017**, *12*, e0178819. [CrossRef]

29. Kim, C.H.; Park, J.H.; Kim, H.; Chung, S.; Park, S.H. Modeling the human body shape in bioimpedance vector measurements. In Proceedings of the 2010 Annual International Conference of the IEEE Engineering in Medicine and Biology, Buenos Aires, Argentina, 31 August–4 September 2010; pp. 3872–3874.

30. Armstrong, L.E.; Pumerantz, A.C.; Fiala, K.A.; Roti, M.W.; Kavouras, S.A.; Casa, D.J.; Maresh, C.M. Human hydration indices: Acute and longitudinal reference values. *Int. J. Sport Nutr. Exerc. Metab.* **2010**, *20*, 145–153. [CrossRef]

31. Lohman, T.G.; Roche, A.F.; Martorell, R. *Anthropometric Standardization Reference Manual*; Human Kinetics Books: Champain, IL, USA, 1988.

32. Frisancho, A.R. Anthropometric standards. In *An Interactive Nutritional Reference of Body Size and Body Composition for Children and Adults*; University of Michigan Press: Ann Arbor, MI, USA, 2008.

33. Lukaski, H.C.; Piccoli, A. Bioelectrical impedance vector analysis for assessment of hydration in physiological states and clinical conditions. In *Handbook of Anthropometry*; Preedy, V., Ed.; Springer: Berlin, Germany, 2012; pp. 287–305.

34. Carter, J.E.L.; Heath, B.H. The heath-carter somatotype rating form. In *The Heath-Carter Anthropometric Somatotype*; San Diego State University: San Diego, CA, USA, 1990.

35. Cheng, H.L.; O'Connor, H.; Kay, S.; Cook, R.; Parker, H.; Orr, R. Anthropometric characteristics of Australian junior representative rugby league players. *J. Sci. Med. Sport* **2014**, *17*, 546–551. [CrossRef]

36. Hohenauer, E.; Rucker, A.M.; Clarys, P.; Küng, U.M.; Stoop, R.; Clijsen, R. Anthropometric and performance characteristics of the German rugby union 7s team. *J. Sports Med. Phys. Fit.* **2017**, *57*, 1633–1641.

37. Di Vincenzo, O.; Marra, M.; Scalfi, L. Bioelectrical impedance phase angle in sport: A systematic review. *J. Int. Soc. Sports Nutr.* **2019**, *16*, 49. [CrossRef]

38. Gutnik, B.; Zuoza, A.; Zuozienė, I.; Alekrinskis, A.; Nash, D.; Scherbina, S. Body physique and dominant somatotype in elite and low-profile athletes with different specializations. *Medicina (Kaunas)* **2015**, *51*, 247–252. [CrossRef]

39. Slimani, M.; Nikolaidis, P.T. Anthropometric and physiological characteristics of male soccer players according to their competitive level, playing position and age group: A systematic review. *J. Sports Med. Phys. Fit.* **2019**, *59*, 141–163. [CrossRef] [PubMed]

40. Cárdenas-Fernández, V.; Chinchilla-Minguet, J.L.; Castillo-Rodríguez, A. Somatotype and body composition in young soccer players according to the playing position and sport success. *J. Strength Cond. Res.* **2019**, *33*, 1904–1911. [CrossRef] [PubMed]

41. Giorgi, A.; Vicini, M.; Pollastri, L.; Lombardi, E.; Magni, E.; Andreazzoli, A.; Orsini, M.; Bonifazi, M.; Lukaski, H.; Gatterer, H. Bioimpedance patterns and bioelectrical impedance vector analysis (BIVA) of road cyclists. *J. Sports Sci.* **2018**, *36*, 2608–2613. [CrossRef] [PubMed]

42. Gatterer, H.; Schenk, K.; Laninschegg, L.; Lukaski, H.; Burtscher, M. Bioimpedance identifies body FluidLoss after exercise in the heat: A pilot study with body cooling. *PLoS ONE* **2014**, *9*, e109729. [CrossRef]

43. Campa, F.; Piras, A.; Raffi, M.; Trofè, A.; Perazzolo, M.; Mascherini, G.; Toselli, S. The effects of dehydration on metabolic and neuromuscular functionality during cycling. *Int. J. Environ. Res. Public Health* **2020**, *17*, 1161. [CrossRef]

44. Pollastri, L.; Lanfranconi, F.; Tredici, G.; Schenk, K.; Burtscher, M.; Gatterer, H. Body fluid status and physical demand during the Giro d'Italia. *Res. Sports Med.* **2016**, *24*, 30–38. [CrossRef]

45. Nescolarde, L.; Yanguas, J.; Lukaski, H.; Alomar, X.; Rosell-Ferrer, J.; Rodas, G. Effects of muscle injury severity on localized bioimpedance measurements. *Physiol. Meas.* **2015**, *36*, 27–42. [CrossRef]

46. Mascherini, G.; Gatterer, H.; Lukaski, H.; Burtscher, M.; Galanti, G. Changes in hydration, body-cell massand endurance performance of professional soccer players through a competitive season. *J. Sports Med. Phys. Fit.* **2015**, *55*, 749–755.

47. Campa, F.; Silva, A.M.; Iannuzzi, V.; Mascherini, G.; Benedetti, L.; Toselli, S. The role of somatic maturation on bioimpedance patterns and body composition in male elite youth soccer players. *Int. J. Environ. Res. Public Health* **2019**, *16*, 4711. [CrossRef]

48. Campa, F.; Gatterer, H.; Lukaski, H.; Toselli, S. Stabilizing bioimpedance-vector-analysis measures with a 10-minute cold shower after running exercise to enable assessment of body hydration. *Int. J. Sports Physiol. Perform.* **2019**, *14*, 1006–1009. [CrossRef]

49. Silva, A.M.; Matias, C.N.; Nunes, C.L.; Santos, D.A.; Marini, E.; Lukaski, H.C.; Sardinha, L.B. Lack of agreement of in vivo raw bioimpedance measurements obtained from two single and multi-frequency bioelectrical impedance devices. *Eur. J. Clin. Nutr.* **2019**, *73*, 1077–1083. [CrossRef] [PubMed]

Article

Are Opera Singers Fit or Not?

Jacopo Junio Valerio Branca [†], Massimo Gulisano *,[†], Mario Marella [†] and Gabriele Mascherini

Department of Experimental and Clinical Medicine, Anatomy and Histology Section, University of Florence, 50134 Florence, Italy; jacopojuniovalerio.branca@unifi.it (J.J.V.B.); mario.marella@unifi.it (M.M.); gabriele.mascherini@unifi.it (G.M.)
* Correspondence: massimo.gulisano@unifi.it; Tel.: +39-055-275-8069
† These Authors contributed equally to the work.

Received: 7 April 2020; Accepted: 16 May 2020; Published: 21 May 2020

Abstract: Little information is available about the physical fitness of opera singers. The aim of this study is to measure cardiac engagement during rehearsals and to test both cardiovascular fitness and body composition in a group of opera singers. Thirty-two opera singers (17 female and 15 male) were enrolled for the assessment of body composition by bio impedance, of cardiovascular fitness by submaximal exercise test on a cycle ergometer and the physical effort during singing. Anthropometric parameters showed an overweight condition mainly due to an increase in fat mass. Watts reached during the cycle ergometer test were below the normal range for the general population. During rehearsals, singers have reached 95% of the maximum heart rate. Nowadays, opera singers show low levels of physical fitness, but singing is an activity that requires a high heart effort. Therefore, it is recommended to involve such professionals in a gradual and individualized physical training program in order to avoid fatigue during performances and achieve a better singing performance.

Keywords: aerobic training; singing; physical fitness; heart rate; physical effort of singing

1. Introduction

In 2008, the New York Times published an article entitled "Singing and Fitness" [1]. This article hypothesized an indirect role of the effects of singing, especially lyrical, on the cardiovascular, pulmonary and hormonal systems. In particular, lyrical singers seem to have some kind of protection against body weight accumulation because the physical stresses caused by their activity promote a strengthening of the chest cavity muscles, a greater pulmonary capacity, better cardiac functionality and an increase in leptin hormone secretion. This article states that even if it has not yet been demonstrated, singing in an opera requires great physical effort due to the continuous moving onstage, sometimes even with heavy costumes. Consequently, being in a good physical condition helps to achieve a better performance by avoiding physical and mental fatigue [1].

The other side of the same coin is knowing the performance, having information about the physical effort required, in order to plan a correct training program. However, to date, the cardiovascular effort to which opera singers are subjected during rehearsals is not yet known and cardiac activity during singing has been monitored only in two studies [2,3]. In particular, in these studies heart activity was associated only with breathing, of which the main outcome was heart rate variability and the aim of these studies was to verify how singing promotes a better state of wellbeing.

In addition, nowadays, the literature has only partially investigated the parameters of physical fitness related to health in this particular population. Therefore, the question posed by the article in the New York Times, "is singing exercise?", is still unanswered. To test the level of opera singers' training, many studies have used respiratory functionality tests. The results are not conclusive, but they show a slight prevalence of higher expiratory functionality in singers when compared to reference values for the general population [4]. When respiratory functionality tests are considered in the context of cardiorespiratory fitness, a study performed on marathon runners compared the same expiratory functionality parameters with the general population without reporting any significant differences; therefore, these assessments appear to provide only inconclusive information [5].

A recent preliminary study of Ksinopoulou et al. [6] reports for the first time the parameters of cardiovascular fitness in opera singers. An incremental cycle ergometer test was performed: the results for maximum oxygen consumption showed parameters lower than the normal range for the general population matched for age and sex. However, to the authors' knowledge, this is the only study that has studied the cardiovascular fitness of this population.

Another important parameter of physical fitness related to health is body composition. Currently, studies on body composition values in opera singers have not yet been carried out; usually studies on this population report only a description of the sample with anthropometric parameters such as height and weight, without any description of the different body compartment.

Since there are currently no published papers on the physical fitness of opera singers, the aim of this study is to report, for the first time, the cardiorespiratory and body composition parameters of an international level group of singers.

2. Materials and Methods

2.1. Participants

Thirty-two opera singers (17 female and 15 male) were enrolled during three consecutive international musical masterclasses from 2015 to 2017 held by Corso d' Opera in Montepulciano (Italy). The singers were from Italy, Ukraine, Korea, the United States and Russia and were professional singers.

All participants provided written consent. The approval of the Human Research and Ethics Committee of the University of Florence was obtained. The study is in accordance with the Declaration of Helsinki.

2.2. Procedures

2.2.1. Body Composition Analysis

The study of body composition attempts to separate and quantify body weight or mass into its basic components. Methodology used for the assessment of body composition was performed by the integration of anthropometry and bioelectrical impedance of the whole body analysis [7].

Anthropometry parameters were weight, height and Body Mass Index (BMI, kg/m^2). In addition, waist circumference was taken with a non-extendable metric tape, flexible and accurate (Holtain Limited, 1.5 m Flexible Tape) [8].

Whole-body bioelectrical impedance analysis (BIA) is an analysis of the resistance generated in soft tissues as an opposition to the flow of an injected alternate current and it is measured by skin electrodes at the hand and foot level (BIA 101 Sport Edition, Akern, Florence, Italy). Resistance (RZ, Ω) is the opposition to the flow of an injected alternating current, at any current frequency, through intra- and extracellular ionic solutions, while reactance (XC, Ω) is the dielectric or capacitive component of cell membranes and organelles, and tissue interfaces: therefore, changes in impedance measurements reflect changes in hydration and cell mass. Phase Angle (PA; in degrees) is the ratio between RZ and XC and represents the relationship between intra- and extracellular volumes. In addition, starting from RZ and XC parameters we estimate the following body compartments parameters: fat free mass (FFM in kg/m), body cellular mass (BCM in kg/m), fat mass (FM in kg/m), total body water (TBW in L/m) and extracellular water (ECW, % TBW) [9].

2.2.2. Cardiorespiratory Fitness

The evaluation of cardiovascular fitness was carried out with a submaximal test by a cycle ergometer, while the evaluation of respiratory functionality was made by spirometry.

Each subject performed an incremental submaximal test on the cycle ergometer. Before starting the test, the heart rate corresponding to 75% of the predicted maximum was checked for each subject [10]. After wearing a heart rate monitor (Zephyr Technology, Annapolis, MD, USA), the subject started pedalling with a resistance of 20 watts at the baseline, maintaining a pedalling frequency of 50–60 rpm. Then, after every minute, the pedalling load was increased of 20 watts: the test terminated when the heart rate achieved the previously calculated value (75% of the predicted maximum). This test was repeated three times with two minutes of recovery time between each set in order to also evaluate the ability to recover from physical effort. Moreover, this parameter is useful for understanding the level of fitness. The duration (seconds) and the distance traveled (meters) was recorded for each test [11].

The tests of respiratory functionality were performed by a portable spirometer (Med Graphics, CPX Express) to achieve a forced expiratory volume in one second (FEV1), a forced vital capacity (FVC), mean forced expiratory flow (FEF) during the middle half of the FVC (FEF 25–75%) and the percentage of FEV1/FVC. The best of at least two tests, reproducible to within 5%, was accepted. The spirometry was repeated at least three times or until duplicate values were obtained which agreed within 5 percent, and the best reproducible values of FEV1 (percentage predicted) and FVC (percentage predicted) were used in the analysis [12]. The FEF 25–75% accepted was obtained from the single test with the greatest sum of FEV and FVC. Values gained from the tests of pulmonary functionality were reported as a percentage of the predicted value. The predicted values of Miller (1986) were used [13].

2.2.3. Physical Effort during Singing

During a dress rehearsal, the physical effort of opera singers was assessed. Before starting the rehearsals, each singer put on a heart rate monitor (Zephyr Technology, Annapolis, MD, USA) at the chest level that was taken off at the first break. The data on a monitored performance lasting 10 minutes were downloaded. The maximum heart rate (HR max), the maximum respiratory rate (RR max) and the maximum body temperature (BT max) were analyzed.

2.2.4. Theory/Calculation

Data are expressed as mean and standard deviation. The body composition values were compared for age and sex of the general population [14]. To test the level of fatigue of opera singers between the three tests at the cycle ergometer an ANOVA test was performed. In addition, cardiovascular fitness data were compared between males and females with the two-tailed Student's *t*-test to evaluate if the performances of the two genders were different [13]. The respiratory functionality tests were compared with the reference values for age and sex. In order to verify the presence of a significant difference of opera singers compared to the predicted values, a paired sample Student's *t*-test was performed. The parameters of maximum heart rate, maximum respiratory rate and maximum body temperature detected during the singing dress rehearsal were compared with reference to the standard physiological parameters.

3. Results

The results on body composition are shown in Table 1. Body composition parameters also show a group with good cellular hydration and health but with an increased fat mass compartment (FM: Female = 20.8 ± 4.4; Male = 11.8 ± 3.2 kg/m).

Table 1. Anthropometrics and body composition parameters of opera singers enrolled.

	Female	Normal Range	Male	Normal Range
Age (yrs)	28.1 ± 4.6		34.1 ± 6.2	
Height (cm)	166.5 ± 6.3		177.3 ± 6.7	
Weight (kg)	69.4 ± 16.1		86.3 ± 6.1	
BMI (kg/m^2)	25.1 ± 6.4	18.5–25	28.8 ± 3.2	18.5–25
Waist c. (cm)	84.5 ± 11.4	< 88	95.8 ± 13.8	< 102
RZ (Ω)	555.2 ± 96.7		432.9 ± 27.2	
XC (Ω)	60.8 ± 8.3		54.7 ± 8.8	
PA (°)	6.3 ± 0.7	5.8–7.4	7.3 ± 1.1	6.7–8.3
FFM (kg/m)	30.8 ± 3.6	23–28	39.2 ± 2.5	28–35
TBW (L/m)	22.5 ± 2.7	15–22	28.8 ± 2.9	18–26
ECW (%)	44.4 ± 3.1	39–45	40.8 ± 3.9	38–44
BCM (kg/m)	17.1 ± 2.9	10–17	23.2 ± 1.9	14–21
FM (kg/m)	20.8 ± 4.4	7–14	11.8 ± 3.2	4–9

Legend: BMI = Body Mass Index; Waist c. = waist circumference; RZ = resistance of flow injected by bioimpedance analyzer; XC = reactance of flow injected by bioimpedance analyzer; PA = phase angle; FFM = fat-free mass; TBW = total body water; ECW = extracellular water; BCM = body cellular mass; FM = fat mass.

The parameters on cardiovascular fitness recorded on cycle ergometer test show better results in females than males. These parameters with higher results in females were maximum power achieved at the cycle ergometer at 75% of maximum heart rate (Female = 61.3 ± 19.3; Male = 48.9 ± 17.1 watts, $p < 0.05$), the distance covered (Female = 1029.3 ± 300.2 m; Male = 838.2 ± 405.7 m; $p < 0.05$) and the duration (Female = 180.3 ± 72.8; Male = 145.4 ± 72.1 sec; $p < 0.05$). The ANOVA test shows that both the second and third test record statistically lower values than the first test in both genders (Figure 1).

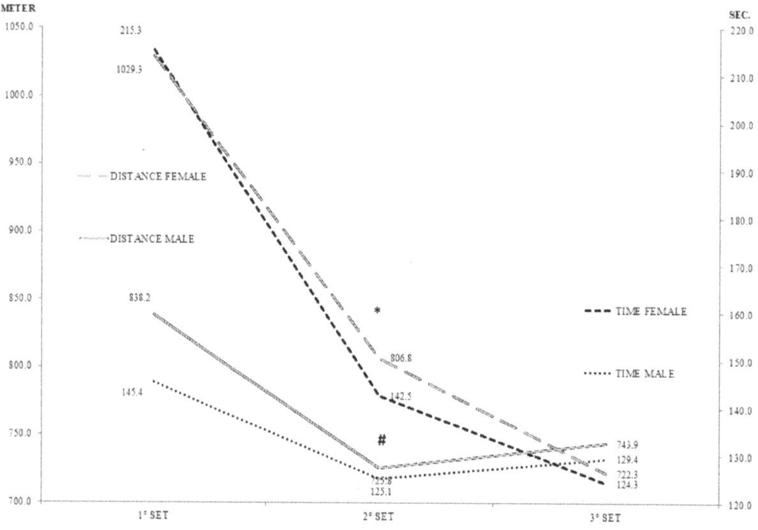

Figure 1. Results of cardiovascular fitness expressed as distance covered and time required during an incremental cycle ergometer test. * statistical difference (ANOVA test) of the distance covered during the three set of cycle ergometer test. # statistical difference (ANOVA test) of the duration during the three set performed on the cycle ergometer test.

Table 2 shows the respiratory functionality results. Both males and females showed normal parameters, without significant differences compared to the values predicted for age and sex.

Table 2. Comparison of respiratory function tests of sample with predicted values.

	Female	Predict	*P* value	Male	Predict	*P* value
FVC (L)	3.9 ± 0.9	4.0	NS	4.7 ± 0.7	5.1	NS
FEV1 (L/sec)	3.3 ± 0.6	3.4	NS	4.1 ± 0.7	4.4	NS
FEV1/FVC (%)	85.1 ± 10.6	84.7	NS	86.8 ± 5.4	82.4	NS
FEF25/75 (L/sec)	3.8 ± 1.3	3.6	NS	4.4 ± 1.3	4.3	NS

Legend: FVC = forced vital capacity; FEV1 = forced expiratory volume in one second; FEV1/FVC = ratio Tiffeneau–Pinelli index; FEF25/75 = mean forced expiratory flow during the middle half of the FVC; FEF Max = the maximum instantaneous flow achieved during a FVC maneuver.

The results of the singing dress rehearsal evaluation are shown in Table 3. All three parameters analyzed showed an increase compared to the physiological values at rest. However, the heart rate increased about 80% compared to the resting values, while the respiratory rate and body temperature showed a lower increase, about 24% and 2.5% respectively.

Table 3. Results of heart rate, respiration rate and body temperature recorded for 10 minutes of an opera singing test. These results were compared with the normal values at rest.

	Female	Values at rest	Male	Values at rest
HR max (bpm)	180.0 ± 8.3	60–100	178.0 ± 13.3	60–100
RR max (bpm)	19.8 ± 3.5	12–16	19.2 ± 4.0	12–16
T max (°)	37.9 ± 0.2	37	37.9 ± 0.1	37

Legend: HR max = maximum heart rate; RR max = maximum respiratory rate; BT max = maximum body temperature.

4. Discussion

Currently, little information is available about the physical effort required during opera singing and the consequent fitness parameters of opera singers. Our sample involved people who had been professional opera singers for more than five years and, therefore, the parameters of physical fitness related to health should be considered stable. The results of the present study on cardiovascular fitness confirmed the preliminary data gained by Ksinopoulou et al. [6] and those obtained on respiratory functionality are in accordance with the literature [15]. However, the present study reported for the first time the body composition parameters of opera singers, both female and male, and their physical effort during singing.

The sample showed anthropometric parameters at the upper limit of the normal range. The analysis of body composition by bioelectrical parameters (in particular the phase angle), despite a high fat mass, reported good cellular health, also confirmed by good cell mass and correct compartmental distribution of body water.

A central (or upper body) pattern of fat distribution is negatively associated with airway function, while increases in body muscular mass result in linear increases for all spirometric variables in healthy persons [16]. The waist circumference and fat-free mass are in the normal range, therefore the results of the spirometric tests can be associated with good functionality. In particular, these results are not different to those predicted for a sample of the general population matched for age and sex.

In order to verify cardiovascular fitness, we assessed the ability to perform a physical activity up to a moderate intensity, as well as the ability to recover from physical effort and then to repeat it. The amount of time needed to reach 75% of the maximum heart rate was 3 minutes for females and 2 minutes and 25 seconds for males. It is therefore clear that opera singers show a low level of cardiovascular fitness, since the maximum watts achieved during an incremental test on the cycle ergometer were lower than the predicted values [17]. Another indicator of a low level of fitness is the difficulty in recovering from an effort and then repeating it; both the second and third tests on the cycle ergometer showed significantly lower values than the first one.

As a result, it may be assumed that singing does not involve a significant heart effort; however, heart rate is the parameter that increases the most during the 10 minutes of monitoring carried out during the dress rehearsal, achieving 95% of the maximum predicted for age and sex. Respiration rate and body temperature increased moderately. The hypothalamus, which is our brain thermostat, regulates the body temperature at 37 °C with the possibility of adjustments of ± 1 °C [18]. Considering that during the performance singers are on a stage with spotlights and heavy clothes, the increase of 1 °C should be considered reasonable. A respiratory rate between 12–16 breaths per minute is considered a resting value [19]. Opera singing has particularly long expiratory phases; it is therefore reasonable to record the respiratory frequencies slightly above the resting values. Therefore, if the results for BT max and RR max can be considered as expected values, the HR max values recorded in rehearsals lead us to conclude that there is a high heart effort for opera singers. Females show better results in both body composition parameters and cardiovascular fitness tests. The sample investigated features a lower age for females than for males, however, the age range is within early adulthood. Therefore, the difference in the aforementioned parameters seems more correlated to lifestyle, such as diet and physical activity performed.

Considering the results for cardiovascular fitness and HR max recorded during the rehearsals, we can state that aerobic training seems both necessary for increasing the level of fitness in opera singers and is functional to the activity they usually perform.

Aerobic exercise, such as walking, jogging, cycling and aerobic gym courses, seems the best training for developing heart-lung strength and endurance. In order to achieve aerobic fitness it is necessary to perform enjoyable exercise and practice training continuously for 30 to 50 minutes, three to five times per week, at 60% to 80% of your maximum heart rate, as suggested by the guidelines of the American College of Sports Medicine [20].

In order to improve physical fitness, the guidelines also recommend including resistance training [21]. Muscular conditioning is an important component in maximizing health but a correct technique in all exercises should be carefully followed in order to avoid bothersome neck or shoulder tension [22]. A resistance training program should be focused on the number of repetitions and not on the amount of weight lifted. At the beginning of a resistance training program, weights should be lifted for one session of 10–15 repetitions. Each exercise session should train eight muscle groups two to three times a week [20].

During a long performance, physical fatigue may occur, defined as peripheral, which can firstly present as gasps after having sung using a microphone. The perseverance of peripheral fatigue can lead to central fatigue, with a consequent loss of decision-making ability and readiness, thus compromising the performance definitively [23]. To avoid this, it is necessary to achieve and maintain good physical fitness, which enables the singer to deal with the high cardiovascular effort required for opera performances.

The present study shows some strengths. Firstly, nowadays it is the only study evaluating cardiac engagement during a rehearsal of opera singers; moreover, assessments were made to establish the level of physical fitness, including body composition. Secondly, the number of subjects who have undergone the assessments is in line with current studies. Finally, the homogeneity of the sample is an advantage; indeed, all the subjects performed singing regularly at an international level.

On the other hand, a limitation of the study is the use of a submaximal test to evaluate cardiorespiratory fitness. However, this was chosen to ensure the safety of the singers, as the study was not carried out in a clinical setting because the evaluation was not for health reasons. In addition, the sample is composed of singers of different races, which could limit the generalization of the results; however, this allowed us to obtain this sample size with such homogeneity with respect to the international level of performance.

The results obtained entail new directions of study in this field. The first one could be to verify the effectiveness, in terms of singing performance, of an exercise program for this specific population. Subsequently, it is possible to check whether any improvements are more sensitive to training in muscle strength or aerobic endurance.

5. Conclusions

The combination of vigorous singing with intense physical recitation requires good physical fitness. In addition, roles that are more active require additional energy and motor skills. Since singers act in different types of productions, often physical charm is also required. Nowadays, the public and producers demand singers who are similar to the characters they perform, thus, it is difficult for an overweight actor to get the role of Butterfly. Having good physical fitness has become a key element for a successful career as an opera singer.

Author Contributions: Conceptualization and data curation, M.M.; Formal analysis, visualization and writing—original draft, G.M.; supervision and validation M.G.; writing—review & editing, J.J.V.B. All authors have read and agreed to the published version of the manuscript.

Funding: This research received no external funding.

Conflicts of Interest: The authors declare no conflict of interest.

References

1. Claiborne Ray, C. Singing and fitness. *New York Times*, 1 April 2008. Available online: http://www.nytimes.com/2008/04/01/science/01qna.html (accessed on 4 March 2020).
2. Vickhoff, B.; Malmgren, H.; Aström, R.; Nyberg, G.; Ekström, S.R.; Engwall, M.; Snygg, J.; Nilsson, M.; Jörnsten, R. Music structure determines heart rate variability of singers. *Front. Psychol.* **2013**, *4*, 334. [CrossRef] [PubMed]
3. Grape, C.; Sandgren, M.; Hansson, L.O.; Ericson, M.; Theorell, T. Does singing promote well-being?: An empirical study of professional and amateur singers during a singing lesson. *Integr. Physiol. Behav. Sci.* **2003**, *38*, 65–74. [CrossRef] [PubMed]
4. Schorr-Lesnick, B.; Teirstein, A.S.; Brown, L.K.; Miller, A. Pulmonary function in singers and wind-instrument players. *Chest* **1985**, *88*, 201–205. [CrossRef] [PubMed]
5. Mahler, D.A.; Moritz, E.D.; Loke, J. Ventilatory responses at rest and during exercise in marathon runners. *J. Appl. Physiol.* **1982**, *52*, 388–392. [CrossRef] [PubMed]
6. Ksinopoulou, H.; Hatzoglou, C.; Daniil, Z.; Gourgoulianis, K.; Karetsi, H. Ergospirometry Findings in Wind Instrument Players and Opera Singers. *Int. J. Occup. Environ. Med.* **2017**, *8*, 60–61. [CrossRef] [PubMed]
7. Mascherini, G.; Petri, C.; Galanti, G. Integrated total body composition and localized fat-free mass assessment. *Sport Sci. Health* **2015**, *11*, 217–225. [CrossRef]
8. Marfell-Jones, M.J.; Stewart, A.D.; de Ridder, J.H. *International Standards for Anthropometric Assessment*; International Society for the Advancement of Kinanthropometry: Wellington, New Zealand, 2012.
9. Kyle, U.G.; Bosaeus, I.; De Lorenzo, A.D.; Deurenberg, P.; Elia, M.; Manuel Gómez, J.; Lilienthal Heitmann, B.; Kent-Smith, L.; Melchior, J.C.; Pirlich, M.; et al. ESPEN, Bioelectrical impedance analysis—Part II: Utilization in clinical practice. *Clin. Nutr.* **2004**, *23*, 1430–1453. [CrossRef] [PubMed]
10. Tanaka, H.; Monahan, K.D.; Seals, D.R. Age-predicted maximal heart rate revisited. *J. Am. Coll. Cardiol.* **2001**, *37*, 153–156. [CrossRef]
11. Fletcher, G.F.; Ades, P.A.; Kligfield, P.; Arena, R.; Balady, G.J.; Bittner, V.A.; Coke, L.A.; Fleg, J.L.; Forman, D.E.; Gerber, T.C.; et al. Exercise standards for testing and training: A scientific statement from the American Heart Association. *Circulation* **2013**, *128*, 873–934. [CrossRef] [PubMed]

12. Redlich, C.A.; Tarlo, S.M.; Hankinson, J.L.; Townsend, M.C.; Eschenbacher, W.L.; Von Essen, S.G.; Sigsgaard, T.; Weissman, D.N. American Thoracic Society Committee on Spirometry in the Occupational Setting. Official American Thoracic Society technical standards: Spirometry in the occupational setting. *Am. J. Respir. Crit. Care Med.* **2014**, *189*, 983–993. [CrossRef] [PubMed]

13. Miller, A. *Pulmonary Function Test in Clinical and Occupational Disease*; Grune & Stratton: Philadelphia, PA, USA, 1986.

14. Barbosa-Silva, M.C.; Barros, A.J.; Wang, J.; Heymsfield, S.B.; Pierson, R.N., Jr. Bioelectrical impedance analysis: Population reference values for phase angle by age and sex. *Am. J. Clin. Nutr.* **2005**, *82*, 49–52. [CrossRef] [PubMed]

15. Thompson, B.R.; Johns, D.P.; Bailey, M.; Raven, J.; Walters, E.H.; Abramson, M.J. Prediction equations for single breath diffusing capacity (Tlco) in a middle aged caucasian population. *Thorax* **2008**, *63*, 889–893. [CrossRef] [PubMed]

16. Maiolo, C.; Mohamed, E.I.; Carbonelli, M.G. Body composition and respiratory function. *Acta Diabetol.* **2003**, *40*, S32–S38. [CrossRef] [PubMed]

17. Usaj, A.; Kandare, F. The oxygen uptake threshold during incremental exercise test. *Pflugers Arch.* **2000**, *440*, R200–R201. [CrossRef] [PubMed]

18. Obermeyer, Z.; Samra, J.K.; Mullainathan, S. Individual differences in normal body temperature: Longitudinal big data analysis of patient records. *BMJ* **2017**, *359*, j5468. [CrossRef] [PubMed]

19. Barrett, K.E.; Ganong, W.F. *Ganong's Review of Medical Physiology*, 25th ed.; McGraw-Hill Medical: New York, NY, USA, 2012; p. 619.

20. Franklin, B.A.; Whaley, M.H. *ACSM's Guidelines for Exercise Testing and Prescription*, 6th ed.; Lippincott Williams & Wilkins: Philadelphia, PA, USA, 2000.

21. McArdle, W.; Katch, F.I.; Katch, V.L. *Exercise Physiology: Nutrition, Energy, and Human Performance*, 8th ed.; Wolters Kluwer Health: Riverwoods, IL, USA, 2014.

22. Eller, N.; Skylv, G.; Ostri, B.; Dahlin, E.; Suadicani, P.; Gyntelberg, F. Health and lifestyle characteristics of professional singers and instrumentalists. *Occup. Med.* **1992**, *42*, 89–92. [CrossRef] [PubMed]

23. Amann, M. Central and peripheral fatigue: Interaction during cycling exercise in humans. *Med. Sci. Sports Exerc.* **2011**, *43*, 2039–2045. [CrossRef] [PubMed]

Article

Higher Levels of Physical Fitness Are Associated with Lower Peak Plantar Pressures in Older Women

Lovro Štefan [1,*], Mario Kasović [1,2] and Martin Zvonař [2,3]

[1] Department of General and Applied Kinesiology, Faculty of Kinesiology, University of Zagreb, 10 000 Zagreb, Croatia; mario.kasovic@kif.hr

[2] Department of Sport Motorics and Methodology in Kinanthropology, Faculty of Sports Studies, Masaryk University, 625 00 Brno, Czech Republic; zvonar@fsps.muni.cz

[3] Faculty of Science, Masaryk University, 625 00 Brno, Czech Republic

* Correspondence: lovro.stefan1510@gmail.com

Received: 8 April 2020; Accepted: 16 May 2020; Published: 18 May 2020

Abstract: Little is known about how physical fitness is associated with peak plantar pressures in older adults. Therefore, the main purpose of the study was to explore whether higher physical fitness levels were associated with lower peak plantar pressures in a sample of community-dwelling older adults. In this cross-sectional study, we recruited 120 older women aged ≥60 years. To assess the level of peak plantar pressure, we used a Zebris plantar pressure platform. To estimate the level of physical fitness, a senior fitness test battery was used. To calculate the associations between the level of physical fitness and peak plantar pressures beneath the different foot regions (forefoot, midfoot and hindfoot), we used generalized estimating equations with a linear regression model. In unadjusted models, higher physical fitness levels were associated with lower peak plantar pressures. When we adjusted for chronological age, the risk of falls and the presence of foot pain, higher physical fitness levels remained associated with lower peak plantar pressures. Our study shows that higher levels of physical fitness are associated with lower peak plantar pressures, even after adjusting for several potential covariates.

Keywords: older adults; exercise; correlation; foot; biomechanics

---

1. Introduction

The percentage of older adults aged ≥65 years has risen dramatically in the last 50 years [1]. It is estimated that the number of older adults will significantly increase in the next 30 years [1]. During the aging process, older adults suffer from a higher prevalence of chronic diseases [2], steadily losing the ability to perform activities in everyday living. Additionally, the number of disabilities doubles and physical limitations quadruples after the age of 60 [3].

The most important factor of aging is being independent and maintaining a high life quality [4]. However, age-related loss of muscle mass and greater accumulation of fat mass often lead to a significant decline in physical performance [5]. Previous evidence has shown that poor physical performance is associated with several health-related consequences, including higher risk of falls [6], higher incidence of chronic diseases [7] and mortality [8]. Physical performance is often associated with physical fitness [9], the capacity needed to undertake daily activities [9]. The level of physical fitness tends to reduce during aging, in terms of strength, endurance, flexibility and agility [10,11]. Such a decrease is often accompanied by a skeletal muscle and joint motion loss and an increment of fat mass [9].

It has been well-documented that elderly individuals have altered foot characteristics, including higher plantar pressures during walking [12]. Such a condition may cause lower extremity pain and discomfort [13], tending to discourage these individuals from achieving higher physical fitness levels. Measuring peak plantar pressure beneath different foot regions is of significant clinical importance, because it is postulated that older adults with abnormal foot posture and balance performance may be of higher risk of generating higher peak plantar pressure loads [14]. One previous study has shown that total physical activity and time spent in moderate-to-vigorous physical activity are significantly and inversely correlated with peak plantar pressures across the heel region of the foot [15]. Another study using similar methodology has shown similar results, i.e., moderate-intensity, vigorous-intensity and moderate- to vigorous-intensity physical activity are significantly and inversely correlated with peak plantar pressures across the middle and lateral forefoot region and the lateral midfoot region of the foot [16]. However, both studies were conducted among children and used objectively measured physical activity as an outcome variable. Evidence shows that older adults suffer from an increased risk of osteoarthritis, myopathies and pain incidence, which often affect walking ability through a reduced muscle strength and endurance [17]. These individuals with lower levels of muscle mass and impaired muscle strength are unable to perform everyday activities and have deviated walking abilities, leading to increased peak plantar pressures [18]. Although physical activity and physical fitness are often closely related, they have been moderately correlated with at times great variability [19]. Thus, the correlations between physical activity and physical fitness and peak plantar pressures may not be the same.

We found no study into available literature that has systematically explored the associations between the different components of physical fitness and peak plantar pressures under the different foot regions. Therefore, the main purpose of the study was to explore whether higher physical fitness levels were associated with lower peak plantar pressures in a sample of community-dwelling older adults. We hypothesized that higher physical fitness levels would be associated with lower peak plantar pressure beneath the different foot regions.

2. Materials and Methods

2.1. Participants

This is a cross-sectional study in which older adults aged ≥60 years from five neighborhoods in the city of Zagreb were recruited. The study protocol and sample size collection are described elsewhere [20]. First, we introduced the study methodology via posters. Of an estimated population of 1500 adults aged ≥60 years living in five neighborhoods, the estimated sample size for the confidence level of 95% and confidence interval of 10% was 110. In order to correct for possible missing data, we recruited 210 participants, of whom 73 did not provide full data and 17 could no longer be in the study due to personal issues. Finally, we based our study on 120 older women (100%). The inclusion criteria were: (1) being ≥60 years old; (2) living independently in the community; (3) passing the short portable mental status questionnaire; (4) being able to ambulate for at least 10 m with or without an aid; (5) being free from neurological diseases; and (6) could arrange their own transport to a testing venue in their community. Before the study began, all participants gave written informed consent. All procedures performed in this study were anonymous and in accordance with the declaration of Helsinki, and were also approved by the ethical committee of the faculty of kinesiology, University of Zagreb, Croatia (ethical code number: 2019).

2.2. Peak Plantar Pressures

Peak plantar pressures were estimated by using a Zebris plantar pressure platform (FDM; GmbH, Munich, Germany; number of sensors: 11,264; sampling rate: 100 Hz; sensor area: 149 cm × 54.2 cm). The methodology and protocols for obtaining gait analysis data are described in detail elsewhere [20]. In brief, each participant was instructed to walk at a comfortable speed across the platform without shoes and socks. Additionally, all participants were required to look straight forward, not targeting the pressure platform. When they reached the end of the walkway, they needed to turn 180° around and continue to walk again over the platform. Finally, when they reached the end of the second walkway (trial), they again turned 180° around and walked a final time across the platform to the end of the walkway. After the completion of measurement, the software generated the data regarding peak plantar pressures beneath the different foot regions of forefoot, midfoot and hindfoot.

2.3. Physical Fitness Assessment

To assess the level of physical fitness, we used the senior fitness test [21]. The measurement protocols and the reliability and validity properties are described elsewhere [21]. We included seven tests in this study: (1) height and weight; (2) chair stand in 30 s; (3) arm curl in 30 s; (4) 2-min step test; (5) chair sit-and-reach test; (6) back scratch test; and (7) 8-foot up-and-go test. In addition, body-mass index was used as an indicator of general adiposity. The chair stand in 30 s test was used to assess lower body strength, and in it participants needed to come to a full stand from a seated position with arms folded across the chest. The arm curl in 30 s was the second test, representing a general measure of upper-body strength, and involved counting the number of times a person could curl a hand weight (5 pounds or 2.3 kg for women and 8 pounds or 3.6 kg for men) through a full range of motion. The third test included a person stepping in place and raising their knees to a height halfway between the patella (kneecap) and iliac crest (front hip bone). This test is a measure of aerobic endurance, and the results were expressed in the number of steps taken during the two minutes. Next, the chair sit-and-reach test aimed to assess lower-body flexibility. The test involved sitting at the front edge of a stable chair with one leg extended and the other foot flat on the floor. With hands on top of each other and arms outstretched, the participant reached as far forward as possible toward the toes. The score was expressed in cm (a higher score was better) and was measured three times, where the best score was taken in the model. The purpose of the back-scratch test was to assess upper-body flexibility, particularly shoulder flexibility. The test involved reaching one hand over the shoulder and down the back as far as possible and the other hand around the waist and up the middle of the back as far as possible, trying to bring the fingers of both hands together. The score was expressed in cm (a higher score was better) and was measured three times, where the best score was taken in the model. Finally, the 8-foot up-and-go test was used to assess agility and dynamic balance. The test involved getting up from a seated position and walking as quickly as possible around a cone that is 8 feet (2.4 m) away and returning to the seated position. The test was performed two times and the results were expressed in seconds. Height and weight were objectively measured.

3. Covariates

We used the downtown fall risk index to assess the risk of falls [22], which is a reliable and valid instrument that measures five modules: previous falls, medication, sensory deficits, mental state and gait. The results are then summarized and give the score between 0 and 11. A higher score indicates higher risk of falls [22]. Presence of foot pain was determined by using the same question as in previous studies: "On most days do you have pain, aching, or stiffness in either of your feet?" [23]. Responses were categorized as: "Yes, pain in one foot", "Yes, pain in both feet", "Yes, but unable to detect the side" and "No, no pain in either foot". Finally, the responses "Yes, pain in one foot", "Yes, pain in both feet'" and "Yes, but unable to detect the side" were collapsed into the "foot pain" category vs. the "no foot pain" category. In addition, we asked each participant about their chronological age.

4. Data Analysis

Basic descriptive statistics are presented as mean ± SD or median (25th–75th percentile range) for normally and non-normally distributed variables. Lastly, the presence of foot pain is presented as the percentage. We calculated z-scores for each physical fitness test. To get an overall physical fitness index, we summed all z-scores. To calculate the associations between the level of physical fitness (separate components and overall physical fitness index) and peak plantar pressures beneath the different foot regions (forefoot, midfoot and hindfoot), we used generalized estimating equations with a linear regression model. For each analysis, we calculated unadjusted (model 1) and adjusted (model 2) associations. The associations in model 2 were adjusted for the risk of falls, the presence of foot pain and chronological age. The results are presented as β coefficients with 95% confidence intervals. All analyses were performed in the Statistical Packages for Social Sciences version 23 (SPSS Inc., Chicago, IL, USA) with a statistical significance of $p \leq 0.05$.

5. Results

Basic descriptive statistics of the study participants are presented in Table 1.

Table 1. Basic descriptive statistics of the study participants ($N = 120$).

Study Variables	Mean ± SD
Age (years)	71.01 ± 6.77
Height (cm)	158.92 ± 21.41
Weight (kg)	70.29 ± 12.97
Peak plantar pressure beneath the forefoot (N/cm^2)	46.73 ± 10.65
Peak plantar pressure beneath the midfoot (N/cm^2)	16.94 ± 7.41
Peak plantar pressure beneath the hindfoot (N/cm^2)	31.57 ± 7.44
Body-mass index (kg/m^2)	26.79 ± 4.42
Chair stand in 30 s (#)	16.53 ± 4.14
Arm curl in 30 s (#)	19.35 ± 4.68
2-min step test (#)	170.36 ± 43.70
Chair sit–and–reach test (cm) *	7.00 (1.00–11.00)
Back scratch test (cm) *	0.75 (−7.75–4.00)
8 feet up–and–go test (s)	5.37 ± 0.96
Downtown Fall Risk Index *	2.00 (1.00–3.00)
Foot pain (% of "Yes" response) **	54.2

* denotes using median (25th–75th percentile range); ** denotes using percentages; # denotes using the number of repetitions.

The associations between physical fitness and peak plantar pressures under the forefoot region are presented in Table 2. All physical fitness tests were significantly and inversely associated with forefoot peak plantar pressure, except for waist circumference and the back scratch test. In a model adjusted for age, the risk of falls and the presence of foot pain, similar significant and inverse associations remained, excluding the back scratch test and chair sit-and-reach test.

Table 2. The associations between separate components and physical fitness index and peak plantar pressure beneath the forefoot region of the foot in older women ($N = 120$).

Study Variables	Model 1			Model 2 *		
	β Coefficient	95% CI	*p*-Value	β Coefficient	95% CI	*p*-Value
Body-mass index	0.11	−0.29 to 0.56	0.576	0.19	−0.18 to 0.66	0.354
Chair stand in 30 s	−1.22	−1.65 to −0.79	<0.001	−1.02	−1.53 to −0.50	<0.001
Arm curl in 30 s	−1.04	−1.33 to −0.75	<0.001	−0.87	−1.20 to −0.53	<0.001
2-min step test	−0.11	−0.14 to −0.08	<0.001	−0.09	−0.12 to −0.05	<0.001
Chair sit-and-reach test	−0.33	−0.54 to −0.12	0.002	−0.18	−0.41 to −0.05	0.130
Back scratch test	−0.13	−0.32 to 0.07	0.198	0.01	−0.18 to 0.21	0.895
8 feet up-and-go test	3.14	1.17 to 5.11	0.002	2.22	0.08 to 4.35	0.042
Physical fitness index	−1.84	−2.31 to −1.37	<0.001	−1.75	−2.42 to −1.08	<0.001

* adjusted for age, the risk of falls and foot pain. $p < 0.05$.

Table 3 shows the associations between physical fitness and peak plantar pressures under the midfoot region. Specifically, all physical fitness components were significantly and inversely associated with the midfoot peak plantar pressure, except for the 2-min step test and 8 feet up-and-go test. In model 2, the arm curl in 30 s, 2-min step test and 8 feet up-and-go test were not significantly associated with midfoot peak plantar pressure.

Table 3. The associations between separate components and physical fitness index and peak plantar pressure beneath the midfoot region of the foot in older women ($N = 120$).

Study Variables	Model 1			Model 2 *		
	β Coefficient	95% CI	*p*-Value	β Coefficient	95% CI	*p*-Value
Body-mass index	0.37	0.13 to 0.68	0.007	0.39	016 to 0.71	0.006
Chair stand in 30 s	−0.62	−0.94 to −0.30	<0.001	−0.57	−0.88 to −0.25	<0.001
Arm curl in 30 s	−0.38	−0.76 to −0.01	0.046	−0.30	−0.71 to 0.10	0.143
2-min step test	−0.03	−0.06 to 0.01	0.081	−0.01	−0.05 to 0.02	0.354
Chair sit-and-reach test	−0.19	−0.31 to −0.07	0.002	−0.14	−0.28 to −0.01	0.036
Back scratch test	−0.22	−0.37 to −0.08	0.003	−0.19	−0.31 to −0.06	0.004
8 feet up-and-go test	1.09	−0.28 to 2.45	0.120	0.82	−0.63 to 2.27	0.266
Physical fitness index	−0.84	−1.44 to −0.24	0.006	−0.90	−1.62 to −0.17	0.015

* adjusted for age, the risk of falls and foot pain. $p < 0.05$.

Finally, Table 4 shows the associations between physical fitness and peak plantar pressures under the hindfoot region. All physical fitness tests were significantly and inversely associated with the hindfoot peak plantar pressure, excluding waist circumference and the back scratch test. In the adjusted model, only the chair stand in 30 s, chair sit-and-reach test and physical fitness index were significantly and inversely associated with the peak plantar pressure beneath the hindfoot region of the foot.

Table 4. The associations between separate components and physical fitness index and peak plantar pressure beneath the hindfoot region of the foot in older women ($N = 120$).

Study Variables	Model 1			Model 2 *		
	β Coefficient	95% CI	*p*-Value	β Coefficient	95% CI	*p*-Value
Body-mass index	−0.17	−0.53 to 0.15	0.316	−0.15	−0.53 to 0.20	0.422
Chair stand in 30 s	−0.48	−0.67 to −0.28	<0.001	−0.34	−0.60 to −0.08	0.011
Arm curl in 30 s	−0.31	−0.57 to −0.05	0.021	−0.13	−0.47 to 0.21	0.448
2-min step test	−0.03	−0.06 to −0.01	0.046	−0.01	−0.05 to 0.02	0.486
Chair sit-and-reach test	−0.26	−0.36 to −0.16	<0.001	−0.20	−0.31 to −0.09	<0.001
Back scratch test	−0.11	−0.27 to 0.06	0.206	−0.04	−0.18 to 0.11	0.622
8 feet up-and-go test	1.46	0.01 to 2.92	0.048	0.96	−0.48 to 2.39	0.191
Physical fitness index	−0.97	−1.23 to −0.71	<0.001	−0.93	−1.48 to 0.37	<0.001

* adjusted for age, the risk of falls and foot pain. $p < 0.05$.

6. Discussion

The main purpose of the study was to explore whether higher physical fitness levels were associated with lower peak plantar pressures in a sample of community-dwelling older adults. Our main findings are: (1) higher levels of the physical fitness components are associated with lower peak plantar pressures beneath the different foot regions; (2) a higher level of physical fitness index is associated with lower peak plantar pressures beneath the different foot regions; and (3) when the model is adjusted for potential covariates, similar associations between physical fitness and peak plantar pressures remained.

In our study, the overall physical fitness index may serve as a multifactorial component associated with peak plantar pressure, followed by the chair stand in 30 s and arm curl in 30 s tests. Previous evidence has shown that lower muscle masses of upper and lower extremities serve as significant predictors of foot function, including higher prevalence of foot pain and pronation [23]. Moreover, it has been documented that greater fat content in the muscle may lead to diseases associated with poorer physical performance and foot function [24]. One previous study has shown that higher peak plantar pressure beneath the hindfoot and midfoot areas of the foot are associated with lower balance test scores, indicating that proprioception and foot stability are important for predicting foot pressure [14]. Other studies have presented the associations between the level of physical activity and peak plantar pressure in children [15]. Specifically, a study by Mickle et al. [15] has shown that higher total physical activity level and time spent in moderate- to vigorous-intensity physical activity are correlated with lower peak pressures across the heel in boys, while no significant correlations between the aforementioned variables in girls were observed. Another study has shown that higher levels of moderate-intensity, vigorous-intensity and moderate- to vigorous-intensity physical activity are significantly correlated with peak plantar pressures beneath the middle and lateral forefoot regions of the foot, while vigorous-intensity physical activity is only correlated with lateral midfoot region [16]. In adults, higher levels of plantar pressures, especially in the lateral aspect, have been associated with discomfort during walking [13]. Therefore, we speculate that older adults who experienced certain discomfort in the foot region were discouraged from being physically active. Indeed, we strengthened our associations by adjusting for several potential covariates, including previously used question to assess the presence of foot pain [25], and still obtained similar associations between physical fitness and peak plantar pressures. We also found that the participants generated their highest pressures beneath the forefoot and the hindfoot regions. Since the hindfoot region is the first part of the foot to contact the ground during walking and the forefoot region is the last part of the push-off phase, higher plantar pressures seem to occur in those regions more frequently [15,16].

Previous studies have shown that older adults have significant foot deviations, including flatter feet, intrinsic foot muscle weakness, altered plantar pressure loading patterns during walking and reduced plantar tactile sensitivity [12]. Such conditions may alternatively lead to pain, discomfort and higher plantar pressures. Moreover, evidence has also shown that higher cognitive processes that modulate behavior, like attention and executive function, significantly decline with age, affecting normal walking and increasing the risk for falls [26].

It has been well-documented that physical functioning is a powerful preventing factor of a number of health conditions in a population of older adults [27]. However, only one fourth of older adults meet the recommended levels of physical activity prescribed by the World Health Organization [28]. Special interventions and health-related policies designed to increase the level of physical activity and fitness through health professionals or friends and family should be implemented within the community where they live.

This study has several limitations. First, a cross-sectional design cannot allow us to conclude about the causality of the association, that is whether lower levels of peak plantar pressures led to higher levels of physical fitness or vice versa. Second, a small sample might have underpowered the associations. Third, we based our study on a sample living in the urban part of the country, speaking Croatian and only of the white race. Fourth, we studied relatively healthy older women and the associations between the level of physical fitness and peak plantar pressures might be differently relevant for less healthy older adults. Fifth, our study only included women, and by including men, correlations might have been different, and we could make the generalizability for both sexes. Finally, we did not collect additional information regarding lifestyle factors, which may mediate or moderate the associations between physical fitness and peak plantar pressure in older individuals. Therefore, the aforementioned limitations should be overcome in future studies aiming to detect the associations between physical fitness and plantar pressure in older adults.

7. Conclusions

Our study shows that higher levels of physical fitness are associated with lower peak plantar pressures beneath the different foot regions of the foot in older women. Although a cross-sectional design of the study has been acknowledged as a limitation, our findings suggest that higher physical fitness level may be a significant predictor of peak plantar pressures among community-dwelling older women.

Author Contributions: Conceptualization, L.Š. and M.K.; data curation, L.Š.; formal analysis, L.Š.; funding acquisition, M.K.; investigation, L.Š.; methodology, L.Š. and M.K.; project administration, M.K.; resources, M.K.; Software, L.Š.; supervision, M.K.; validation, L.Š.; visualization, L.Š.; writing—original draft, L.Š., M.K. and M.Z.; writing—review and editing, L.Š., M.K. and M.Z. All authors have read and agreed to the published version of the manuscript.

Funding: This research received no external funding.

Acknowledgments: We would like to thank all the participants for their enthusiastic participation in the study.

Conflicts of Interest: The authors declare no conflict of interest.

References

1. United Nations. *World Population Prospects: The 2004 Revision*; UN: New York, NY, USA, 2005.
2. Fong, J.H. Disability incidence and functional decline among older adults with major chronic diseases. *BMC Geriatr.* **2019**, *19*, 323. [CrossRef]
3. Rimmer, J.H. *Fitness and Rehabilitation Programs for Special Populations*; Brown Benchmark: Madison, WI, USA, 1994.
4. Ciprandi, D.; Zago, M.; Bertozzi, F.; Sforza, C.; Galvani, C. Influence of energy cost and physical fitness on the preferred walking speed and gait variability in elderly women. *J. Electromyogr. Kinesiol.* **2018**, *43*, 1–6. [CrossRef]
5. Nygård, L.; Mundal, I.; Dahl, L.; Šaltytė Benth, J.; Rokstad, A. Nutrition and physical performance in older people-effects of marine protein hydrolysates to prevent decline in physical performance: A randomised controlled trial protocol. *BMJ Open* **2018**, *8*, e023845. [CrossRef]
6. Singh, D.K.A.; Pillai, S.G.K.; Tan, S.T.; Tai, C.C.; Shahar, S. Association between physiological falls risk and physical performance tests among community-dwelling older adults. *Clin. Interv. Aging* **2015**, *10*, 1319–1326. [CrossRef]
7. Booth, F.W.; Roberts, C.K. Linking performance and chronic disease risk: Indices of physical performance are surrogates for health. *Br. J. Sports Med.* **2008**, *42*, 950–952. [CrossRef]
8. Veronese, N.; Stubbs, B.; Fontana, L.; Trevisan, C.; Boltezza, F.; Rui, M.; Sartori, L.; Musacchio, E.; Zambon, S.; Maggi, S.; et al. A comparison of objective physical performance tests and future mortality in the elderly people. *J. Gerontol. A Biol. Sci. Med. Sci.* **2017**, *72*, 362–368. [CrossRef]
9. Kostić, R.; Pantelić, S.; Uzunović, S.; Djuraskovic, R. A comparative analysis of the indicators of the functional fitness of the elderly. *Facta Univ. Ser. Phys. Educ. Sport* **2011**, *9*, 161–171.
10. Riebe, D.; Blissmer, B.J.; Greaney, M.L.; Garber, C.E.; Lees, F.D.; Clark, P.G. The relationship between obesity, physical activity, and physical function in older adults. *J. Aging Health* **2009**, *21*, 1159–1178. [CrossRef]
11. Tuna, H.D.; Edeer, A.O.; Malkoc, M.; Aksakoglu, G. Effect of age and physical activity level on functional fitness in older adults. *Eur. Rev. Aging Phys. Act.* **2009**, *6*, 99–106. [CrossRef]
12. Scott, G.; Menz, H.B.; Newcombe, L. Age-related differences in foot structure and function. *Gait Posture* **2007**, *26*, 68–75. [CrossRef]
13. Menz, H.B.; Fotoohabadi, M.R.; Munteanu, S.E.; Zammit, G.V.; Gilheany, M.F. Plantar pressure and relative metatarsal lengths in older people with and without forefoot pain. *J. Orthop. Res.* **2013**, *31*, 427–433. [CrossRef]
14. Mickle, K.J.; Cliff, D.P.; Munro, B.J.; Okely, A.D.; Steele, J.R. Relationship between plantar pressures, physical activity and sedentariness among preschool children. *J. Sci. Med. Sports* **2011**, *14*, 36–41. [CrossRef]
15. Mohd Said, A.; Justine, M.; Manaf, H. Plantar pressure distribution among older persons with different types of foot and its correlation with functional reach distance. *Scientifica (Cairo)* **2016**, *2016*, 8564020. [CrossRef]

16. Riddiford-Harland, D.L.; Steele, J.R.; Cliff, D.P.; Okely, A.D.; Morgan, P.J.; Jones, R.A.; Baur, L.A. Lower activity levels are related to higher plantar pressures in overweight children. *Med. Sci. Sports. Exerc.* **2015**, *47*, 357–362. [CrossRef]

17. Martínez-Vizcaíno, V.; Sánchez-López, M. Relationship between physical activity and physical fitness in children and adolescents. *Rev. Esp. Cardiol.* **2008**, *61*, 108–111. [CrossRef]

18. Lenzi, T.; Carrozza, M.C.; Agrawal, S.K. Powered hip exoskeletons can reduce the user's hip and ankle muscle activations during walking. *IEEE Trans. Neural. Syst. Rehabil. Eng.* **2013**, *21*, 938–948. [CrossRef]

19. Chen, B.; Ma, H.; Qin, L.-Y.; Gao, F.; Chan, K.-M.; Law, S.-W.; Qin, L.; Liao, W.-H. Recent developments and challenges of lower extremity exoskeletons. *J. Orthop. Transl.* **2016**, *5*, 26–37. [CrossRef]

20. Kasović, M.; Štefan, L.; Zvonar, M. Domain-specific and total sedentary behavior associated with gait velocity in older adults: The mediating role of physical fitness. *Int. J. Environ. Res. Public Health* **2020**, *17*, 593. [CrossRef]

21. Rikli, R.E.; Jones, J. Functional fitness normative scores for community-resident older adults, 60–94. *J. Phys. Act. Health* **1999**, *7*, 162–181.

22. Rosendhal, E.; Lundin-Olsson, L.; Kallin, K.; Jensen, J.; Gustafson, Y.; Nyberg, L. Prediction of falls among older people in residential care facilities by the Downton index. *Aging Clin. Exp. Res.* **2003**, *15*, 142–147. [CrossRef]

23. Awale, A.; Hagedorn, T.J.; Dufour, A.B.; Menz, H.B.; Casey, V.A.; Hannan, M.T. Foot function, foot pain, and falls in older adults: The Framingham Foot Study. *Gerontology* **2017**, *63*, 318–324. [CrossRef]

24. Visser, M.; Kritchevsky, S.B.; Goodpaster, B.H.; Newman, A.B.; Stamm, E.; Harris, T.B. Leg muscle mass and composition in relation to lower extremity performance in men and women aged 70 to 79: The health, aging and body composition study. *J. Am. Geriatr. Soc.* **2002**, *50*, 897–904. [CrossRef]

25. Ferrucci, L.; Penninx, B.W.; Leveille, S.G.; Corti, M.C.; Pahor, M.; Wallace, R.; Harris, T.B.; Havlik, R.J.; Guralnik, J.M. Characteristics of nondisabled older persons who perform poorly in objective tests of lower extremity performance. *J. Am. Geriatr. Soc.* **2000**, *48*, 1102–1110. [CrossRef]

26. Verghese, J.; Mahoney, J.; Ambrose, A.F.; Wang, C.; Holtzer, R. Effect of aging remedation on gait in sedentary seniors. *J. Gerontol. A Biol. Sci. Med. Sci.* **2010**, *65*, 1338–1343. [CrossRef]

27. Haskell, W.L.; Blair, S.N.; Hill, J.O. Physical activity: Health outcomes and importance for public health policy. *Prev. Med.* **2009**, *49*, 280–282. [CrossRef]

28. World Health Organization. *Global Recommendations of Physical Activity for Health*; WHO Press: Geneva, Switzerland, 2011.

 *sustainability*

Article

The Impact of the Weight Status on Cardiovascular Parameters Related to Physical Effort in Young Athletes

Gabriele Mascherini [1,*], Cristian Petri [1], Laura Stefani [1], Loira Toncelli [1], Vittorio Bini [2], Piergiuseppe Calà [3] and Giorgio Galanti [1]

[1] Department of Experimental and Clinical Medicine, University of Florence, 50134 Florence, Italy; cristian.petri@unifi.it (C.P.); laura.stefani@unifi.it (L.S.); loira.toncelli@unifi.it (L.T.); giorgio.galanti@unifi.it (G.G.)
[2] Dipartimento di Medicina, Università di Perugia, 06156 Perugia, Italy; vittorio.bini@unipg.it
[3] Sector "Health and safety in the workplace and special processes in the field of prevention", Directorate of Citizenship Rights and Social Cohesion, Tuscany Region, 50139 Florence, Italy; piergiuseppe.cala@regione.toscana.it
* Correspondence: gabriele.mascherini@unifi.it; Tel.: +39-339-689-5925

Received: 26 April 2020; Accepted: 7 May 2020; Published: 12 May 2020

Abstract: Excess weight leads to an impaired cardiovascular response to physical exertion even at a young age. Sports training during youth promotes cardiovascular adaptations. The aim of the study is to verify the impact of weight status on cardiovascular parameters related to physical effort in young people who engage in competitive sports. A retrospective study was conducted on 8307 young athletes (5578 males and 2729 females) aged 6–18 years (mean age 13.9 ± 2.2 years). The data concerning graded exercise tests of young athletes in normal weight and overweight were compared. Approximately, 13.4% of the sample had excess weight. Young overweight athletes show a higher resting heart rate as well as systolic and diastolic pressure than young normal weight athletes. Excess weight condition leads to a reduction in the duration of the graded exercise test, reaching higher blood pressure values at the end of the test compared to those with normal weight. After four min from the end of the test, heart rate and systolic/diastolic blood pressure remained higher in the young overweight athletes. Excess weight affects cardiovascular parameters both at rest and in response to physical exertion during youth; however, competitive sport seems to be able to keep these parameters within the normal range even in young overweight athletes.

Keywords: youth; overweight; competitive sport; exercise test; obese; heart rate; incremental test; blood pressure

1. Introduction

The prevalence of obesity in children and adolescents is a current public health problem [1]. Excess weight involves both metabolic and hemodynamic alterations [2]. The main hemodynamic changes in obese youth are an increase in resting blood pressure and resting heart rate and an increase in peak heart rate during exercise tests compared to lean controls [3].

Cardiovascular parameters related to physical effort are generally assessed through the graded exercise test (GXT) which currently is widely used for detection of coronary artery disease, prediction of cardiovascular events, and evaluation of physical capacity [4]. In normal subjects, during exercise there is an increase in muscle work, which leads to an increase in oxygen demand. To meet these growing requirements, the cardiovascular system applies a gradual rise in cardiac output. [5].

During GXT, heart rate (HR) and arterial blood pressure (BP) are the main parameters monitored. Dynamic exercise involves an HR increase linear with workload and oxygen demand [6]. The speed in reduction of HR after termination of the GXT, termed HR recovery, currently is considered as an index of cardiorespiratory fitness and is associated with mortality risk [7].

In addition for BP, dynamic exercise induces a systolic blood pressure (SBP) increase because of increasing cardiac output, whereas diastolic blood pressure (DBP) remains the same or slightly reduced due to peripheral vasodilatation [8]. After exercise, the decrease of cardiac output produces a decline in systolic blood pressure, usually reaching resting levels within six min [9].

The interpretation of GXT data in pediatric subjects requires special considerations; these results may have different connotations based on the age, size, and sex of the child [10]. The impact of body weight is well established on HR response to exercise in children, particularly during an incremental test, the increase in peripheral oxygen demand increases the HR in direct proportion to body weight [11]. Similarly, age and body mass index (BMI) are related to exercise duration in both boys and girls; age has a direct correlation, while BMI has an inverse relationship [12].

However, the influence of BMI on the recovery process after maximal exercise test, measured with HR recovery, is not clear [7,12,13]. Maximal BP response to exercise in children appear mostly influenced by age and sex [14]. Systolic BP decreased about 10% four min after the end of the test in both sexes from 6 to 11 years [15,16]; this recovery process appear to be influenced by excess body weight in non-athletic children [17]. Currently the influence of body weight on cardiovascular performance in young people has been studied mainly on small samples sizes [11], in non-athletic subjects [13] and using different protocols. In addition, most studies with a large sample size did not examine the role of weight status on cardiovascular response to exercise [12,14–16].

Young people who engage in competitive sports have improvements in physical performance and a better cardiovascular response to effort than non-sports peers [18]. While sports appear to play a role in reducing the prevalence of youth obesity [19], it is unclear whether it also has a protective role for cardiovascular parameters related to physical exertion.

Therefore, the purpose of the study is to verify the impact of weight status on the cardiovascular parameters before, during, and after a graded exercise test in a large population of young athletes.

2. Materials and Methods

Since 1982, all children in Italy have undergone pre-participation screening to obtain eligibility for competitive sports. A retrospective study was conducted on the data coming from surveillance carried out during the pre-participation screening of sports eligibility. These data came from the regional reference center for Sports Medicine of the Tuscany Region, Italy in the period of 1 January 1998 to 31 October 2018, inclusively.

2.1. Study Population

We analyzed the data from 8307 young athletes (5578 males and 2729 females) aged between 6 and 18 years (mean age 13.9 ± 2.2 years). Inclusion criteria for the subjects were: Caucasian, practice sports at a competitive level, and not to have any contraindications to sports eligibility. Exclusion criteria in the analyses were having already carried out the same visit and being already included in the study sample at a lower age, having received a temporary suspension of sports eligibility, having received indications for further clinical and/or instrumental evaluations, having received sports eligibility for less than 12 months and an age outside the range of ±6 months compared to the average age of its own stratum.

The study was carried out in conformity with the ethical standards laid down in the 1975 Declaration of Helsinki. This study is part of a project of the Tuscany Region called "Sports Medicine to support regional surveillance systems". It was approved in the Regional Prevention Plan 2014–2018 with the number O-Range18. All data were processed anonymously.

2.2. Clinical Evaluation

The procedures of clinical evaluation were in accordance with the Italian protocol [20], which includes family and personal history, physical examination, and finally cardiological evaluation.

The physical examination also included the measurement of height and weight by trained personnel, using appropriate equipment (Seca GmbH & Co., Hamburg, Germany). BMI was calculated using the formula weight/height^2 (kg/m^2). In order to define normal weight (NW) and overweight/obese status (O&O), a subdivision according to BMI, age, and sex was adopted following the classification based on the International Obesity Task Force (IOTF) for children and adolescent [21].

Cardiological evaluation consisted of a 12-lead electrocardiogram (ECG) at rest and during a GXT. Appropriate-sized and adjustable equipment was used to meet the demands of the pediatric age. Both sexes performed GXT on treadmills or cycle ergometer. The GXT was performed by increasing the speed and/or the grade of the treadmill or by increasing the work on the cycle ergometer with the same pedaling frequency. In particular, Bruce modified protocol for the treadmill and an increase of 25 W every 3 min with a pedaling frequency of 70 rpm for the cycle ergometer were adopted [4] with up to 85% of age-predicted maximal HR [20].

Sport eligibility certification is 1 year and generally it does not coincide with the beginning of the competitive season. Therefore, the evaluations were performed during the regular season.

2.3. Variables

The variables used in this study derive from the clinical evaluation for eligibility in competitive sports. Those resulting from the physical exam are gender, age, height, weight, BMI, weight status (NW or O&O), resting HR and resting BP. Those resulting from the GXT are the number of completed stages, HR at the interruption of the test (HR max), HR after 4 min post-test (HR rec), systolic and diastolic BP at the end of the test (SBP max, DBP max), and the systolic and diastolic BP after 4 min post-test (SBP rec, DBP rec). In order to evaluate the magnitude of increase and decrease during and after the GXT, the differences between resting, maximum, and recovery values were calculated for heart rate, systolic and diastolic blood pressure (Δ HR max-rest, Δ HR max-rec, Δ SBP max-rest, Δ SBP max-rec, Δ DBP max-rest, Δ DBP max-rec).

2.4. Statistical Analysis

The Mann–Whitney test was used to analyze independent and non-normally distributed continuous variables (deviation from Gaussian distribution was checked by using the Shapiro–Wilks test); data, for simplicity, are shown as mean ± SD. All statistical analyses were performed using IBM-SPSS®version 25.0 (IBM Corp., Armonk, NY, USA, 2017). In all analyses, a two-sided p-value ≤0.05 was considered significant.

3. Results

Starting from 8307 subjects that were included in the study, 3075 males and 1299 females carried out the cycle ergometer GXT and 2503 males and 1430 females carried out the treadmill GXT. The prevalence of overweight and obesity in the whole sample was 13.4% (14.7% of males and 10.8% of females).

In the whole group, the excess weight condition led to higher resting HR (NW = 84.2 ± 14.2; O&O = 86.9 ± 14.4 bpm; $p < 0.001$), SBP (NW = 102.2 ± 11.6; O&O = 106.2 ± 12.7 bpm; $p < 0.001$), and DBP values (NW = 63.9 ± 8.3; O&O = 66.3 ± 8.3; $p < 0.001$).

On average, the O&O subjects managed to perform fewer stages during the GXT (Table 1. This occurs mainly on the tests performed on the treadmill (NW = 7.8 ± 1.0, O&O = 7.4 ± 1.0; $p < 0.001$) and in males (NW = 7.6 ± 1.5, O&O = 7.3 ± 1.4; $p < 0.001$). No differences were found in the maximum HR values, not even based on the sex and type of exercise performed (Figures 1 and 2).

Table 1. Differences in Δ HR and in Δ BP performances according to weight status, gender, and type of exercise performed.

		Age (years.)	BMI (kg/m²)	n. stages	Δ HR max-rest (bpm)	Δ HR max-rec (bpm)	Δ SBP max-rest (mmHg)	Δ DBP max-rest (mmHg)	Δ SBP max-rec. (mmHg)	Δ DBP max-rec. (mmHg)
Whole Group	NW (7190)	14.0 ± 2.2	**19.1 ± 2.2 ****	**7.2 ± 1.5 ****	**95.3 ± 13.8 ****	**78.1 ± 13.1 ***	**34.6 ± 14.7 ***	**10.4 ± 8.7 ****	**28.5 ± 13.1 ***	5.9 ± 8.0
	O&O (1117)	13.5 ± 2.2	24.3 ± 2.7	7.1 ± 1.4	92.9 ± 13.4	77.3 ± 12.1	35.8 ± 15.7	11.5 ± 8.7	29.5 ± 13.9	6.4 ± 7.8
Males	NW (4758)	14.2 ± 2.2	**19.3 ± 2.2 ****	**7.6 ± 1.5 ****	**97.3 ± 13.4 ****	**78.9 ± 12.9 ***	**36.5 ± 15.2 ***	**11.0 ± 8.9 ***	30.1 ± 13.6	6.2 ± 8.0
	O&O (820)	13.7 ± 2.1	24.4 ± 2.8	7.3 ± 1.4	94.5 ± 12.8	77.7 ± 12.2	37.9 ± 16.3	11.9 ± 8.9	31.1± 13.7	6.5 ± 7.9
Females	NW (2431)	13.7 ± 2.2	**18.8 ± 2.2 ****	6.5 ± 1.3	**91.3 ± 13.8 ***	76.5 ± 13.3	**30.7 ± 12.7 ***	9.1 ± 8.2	25.4 ± 11.4	5.3 ± 7.8
	O&O (298)	12.8 ± 2.3	24.0 ± 2.6	6.4 ± 1.2	88.5 ± 14.1	76.1 ± 11.7	29.7 ± 12.2	10.6 ± 8.0	25.0 ± 13.4	6.1 ± 7.6
Cycle	NW (3814)	14.3 ± 2.2	**19.4 ± 2.1 ****	6.8 ± 1.7	**97.0 ± 13.8 ****	**77.2 ± 13.2 ***	**38.1 ± 15.5 ***	10.8 ± 9.0	**30.5 ± 13.9 ***	5.0 ± 8.0
	O&O (560)	13.8 ± 2.1	24.5 ± 2.7	6.8 ± 1.7	94.2 ± 13.2	76.1 ± 12.1	39.6 ± 17.1	11.5 ± 8.9	31.9 ± 14.4	5.2 ± 7.8
Treadmill	NW (3377)	13.7 ± 2.4	**18.9 ± 2.3 ****	**7.8 ± 1.0****	**93.3 ± 13.5 ***	79.1 ± 12.9	**30.6 ± 12.6***	**9.8 ± 8.4 ****	26.2 ± 11.7	7.0 ± 7.8
	O&O (556)	13.2 ± 2.3	24.1 ± 2.7	7.4 ± 1.0	91.7 ± 13.6	78.5 ± 12.0	31.9 ± 13.1	11.5 ± 8.5	27.1 ± 13.0	7.6 ± 7.7

Legend: HR: heart rate; BP: blood pressure; NW: normal weight; O&O: overweight and obese; SBP: systolic blood pressure; DBP: diastolic blood pressure. * $p < 0.05$ between normal weight and overweight and obese subjects, ** $p < 0.001$ between normal weight and overweight and obese subjects. In parentheses, the number of subjects referred to the group.

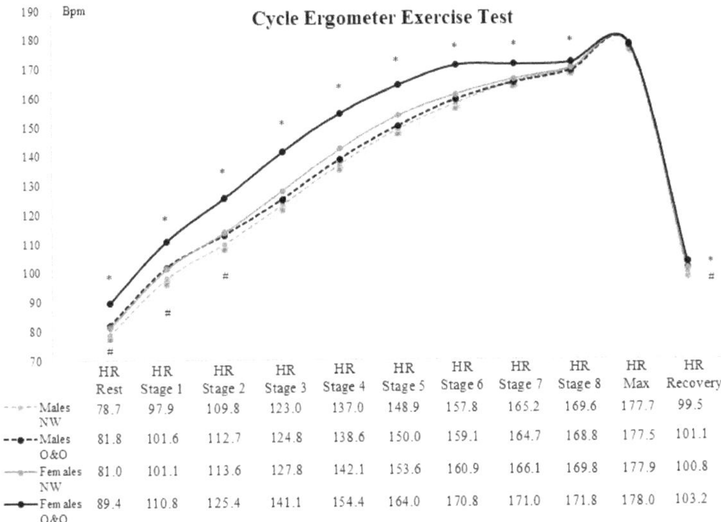

	HR Rest	HR Stage 1	HR Stage 2	HR Stage 3	HR Stage 4	HR Stage 5	HR Stage 6	HR Stage 7	HR Stage 8	HR Max	HR Recovery
Males NW	78.7	97.9	109.8	123.0	137.0	148.9	157.8	165.2	169.6	177.7	99.5
Males O&O	81.8	101.6	112.7	124.8	138.6	150.0	159.1	164.7	168.8	177.5	101.1
Females NW	81.0	101.1	113.6	127.8	142.1	153.6	160.9	166.1	169.8	177.9	100.8
Females O&O	89.4	110.8	125.4	141.1	154.4	164.0	170.8	171.0	171.8	178.0	103.2

Figure 1. Differences in HR performances during cycle ergometer graded exercise test (GXT) according to weight status and gender. Legend: HR: heart rate, NW: normal weight; O&O: overweight and obese. * $p < 0.05$ between normal weight vs. overweight and obese females, # $p < 0.05$ between normal weight vs. overweight and obese males.

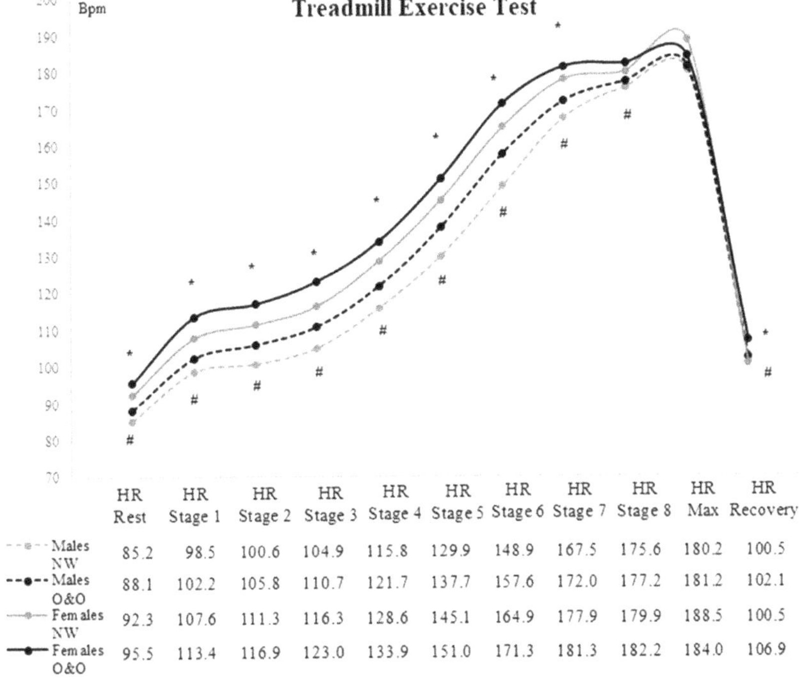

	HR Rest	HR Stage 1	HR Stage 2	HR Stage 3	HR Stage 4	HR Stage 5	HR Stage 6	HR Stage 7	HR Stage 8	HR Max	HR Recovery
Males NW	85.2	98.5	100.6	104.9	115.8	129.9	148.9	167.5	175.6	180.2	100.5
Males O&O	88.1	102.2	105.8	110.7	121.7	137.7	157.6	172.0	177.2	181.2	102.1
Females NW	92.3	107.6	111.3	116.3	128.6	145.1	164.9	177.9	179.9	188.5	100.5
Females O&O	95.5	113.4	116.9	123.0	133.9	151.0	171.3	181.3	182.2	184.0	106.9

Figure 2. Differences in HR performances during treadmill GXT according to weight status and gender. Legend: HR: heart rate, NW: normal weight; O&O: overweight and obese. * $p < 0.05$ between normal weight vs. overweight and obese females, # $p < 0.05$ between normal weight vs. overweight and obese males.

SBP values reached at the end of the test were higher in O&O subjects (NW = 136.7 ± 18.8; O&O = 141.9 ± 20.2 mmHg; $p < 0.001$; Figures 3 and 4). The maximum DBP was on average higher in the O&O (NW = 53.5 ± 8.0; O&O = 54.7 ± 8.2 mmHg; $p < 0.001$); however, this difference disappeared by taking the test on the treadmill (NW = 53.7 ± 7.9, O&O = 54.3 ± 8.2 mmHg; $p > 0.05$; = NS; Figure 4).

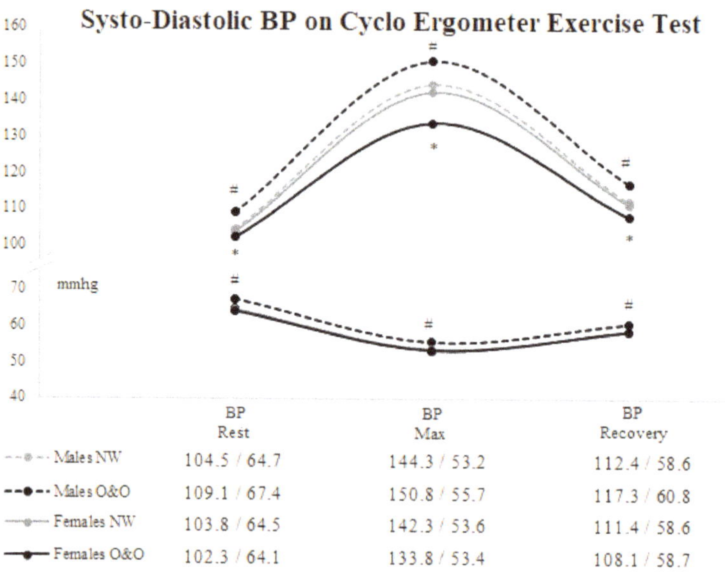

Figure 3. Differences in blood pressure (BP) performances during cycle ergometer GXT according to weight status and gender. Legend: BP: blood pressure; NW: normal weight; O&O: overweight and obese. * $p < 0.05$ between normal weight vs. overweight and obese females, # $p < 0.05$ between normal weight vs. overweight and obese males.

After 4 min from the end of the GXT, in whole group the HR remained higher in the O&O subjects (NW = 101.4 ± 13.1, O&O = 102.6 ± 12.7 bpm; $p < 0.05$; Figures 1 and 2). The same behavior is also described in Figures 3 and 4 for SBP (NW = 108.2 ± 13.4, O&O = 112.4 ± 14.3 mmHg; $p < 0.001$) and for DBP (NW = 59.4 ± 7.8, O&O = 61.1±7.8 mmHg; $p < 0.001$).

The results relating to the magnitude of the increase and decrease in HR, SBP, and DBP are shown in Table 1. These values show a high statistical difference ($p < 0.001$) in ΔHR max-rest and ΔSBP max-rest, a statistical difference ($p < 0.05$) in ΔHR max-rec and ΔSBP max-rec, and finally no difference in the ΔDBP max-rec between athletic subjects NW and O&O.

Figures 1 and 2 show the chronotropic increase and decrease during and after the GXT. In particular, Figure 1 shows that from the 3rd stage there are no differences in HR in male subjects who performed the cycle ergometer GXT (NW = 123.0 ± 18.5, O&O = 124.8 ± 18.0 bpm; p = NS). Table 1 shows that on the cycle ergometer there were no differences on the number of competed stages (NW = 6.8 ± 1.7, O&O = 6.8 ± 1.7 bpm; p = NS) between NW and O&O subjects.

Table 2 reports the results stratified by age and gender. The differences in cardiovascular parameters between NW and O&O appear greater in males and with increasing age. However, the differences already begin in the age group of 8–10 years: resting BP is greater in O&O males. From the age group 12–14 years and onwards, all parameters analyzed (except HR max) show higher values in males. In contrast, O&O athletic females show higher BP values in comparison with NW athletic females.

Table 2. Differences in HR and BP performances at rest, at the maximum effort and after 4 min of recovery according to weight status, age, and gender.

			Rest			Maximum			Recovery		
			HR rest (bpm)	SBP rest (mmhg)	DBP rest (mmhg)	HR max (bpm)	SBP max (mmhg)	DBP max (mmhg)	HR rec. (bpm)	SBP rec. (mmhg)	DBP rec. (mmhg)
8–10 years	Males	NW (176)	86.4 ± 14.2	**88.5 ± 8.0 ***	**57.2 ± 7.0 ***	180.1 ± 8.7	**116.1 ± 14.4 ***	50.2 ± 7.7	93.7 ± 13.3	91.6 ± 10.2 *	55.3 ± 7.2
		O&O (32)	89.4 ± 13.0	93.9 ± 8.3	60.3 ± 5.7	180.7 ± 6.9	122.7 ± 14.2	50.9 ± 8.6	93.9 ± 12.5	95.9 ± 9.5	55.8 ± 8.2
	Females	NW (145)	95.3 ± 13.0	89.7 ± 9.0	**57.2 ± 7.8 ***	183.2 ± 7.6	113.3 ± 11.3	49.8 ± 8.0	100.9 ± 11.5	91.0 ± 9.7 *	54.8 ± 6.5
		O&O (42)	94.8 ± 12.5	92.0 ± 9.9	60.4 ± 7.1	184.9 ± 6.8	117.5 ± 15.0	51.7 ± 7.4	102.7 ± 15.6	95.4 ± 9.8	56.4 ± 5.4
10–12 years	Males	NW (488)	85.6 ± 12.7	**94.8 ± 8.7 ****	**59.8 ± 7.3 ****	180.9 ± 7.0	**124.6 ± 14.4 ****	52.1 ± 7.7	**96.3 ± 13.1 ***	**98.8 ± 10.0 ****	**57.5 ± 7.3 ***
		O&O (108)	88.4 ± 13.6	100.1 ± 10.3	63.1 ± 6.9	181.3 ± 8.5	132.1 ± 15.0	53.1 ± 6.9	99.7 ± 13.6	105.3 ± 10.4	59.4 ± 6.3
	Females	NW (415)	92.5 ± 14.1	**94.1 ± 9.0 ****	**59.2 ± 6.8 ****	183.1 ± 7.1	121.8 ± 13.1	51.1 ± 7.5	104.2 ± 15.0	**97.3 ± 9.7 ***	**56.8 ± 7.1 ***
		O&O (62)	95.4 ± 13.3	99.8 ± 11.7	63.9 ± 7.5	183.8 ± 8.6	124.4 ± 15.3	51.5 ± 7.0	105.1 ± 12.2	103.5 ± 16.6	59.8 ± 7.4
12–14 years	Males	NW (1406)	**83.9 ± 13.2 ****	**99.7 ± 10.3 ****	**62.1 ± 7.7 ****	180.1 ± 7.5	**133. 5 ± 16.1 ****	**54.4 ± 8.4 ****	**99.3 ± 12.7 ***	**105.8 ± 11.9 ****	**58.2 ± 7.1 ****
		O&O (294)	87.1 ± 12.4	105.2 ± 12.4	65.6 ± 7.9	180.3 ± 7.6	139.9 ± 16.7	52.1 ± 7.8	101.5 ± 12.4	110.9 ± 12.1	60.6 ± 7.8
	Females	NW (758)	**92.0 ± 14.3 ***	**99.4 ± 9.8 ****	**62.5 ± 7.2 ***	181.1 ± 7.1	**130.6 ± 14.7 ****	53.2 ± 7.4	104.6 ± 13.6	**104.9 ± 10.9 ****	**58.7 ± 7.6 ***
		O&O (98)	94.5 ± 14.0	103.7 ± 10.6	64.4 ± 7.7	181.6 ± 6.9	134.9 ± 12.4	54.2 ± 6.9	106.7 ± 12.5	107.9 ± 10.7	60.9 ± 6.3
14–16 years	Males	NW (1555)	**80.2 ± 13.0 ***	**107.0 ± 10.6 ****	**66.2 ± 8.1 ****	178.5 ± 8.1	**145.4 ± 16.8 ****	**54.2 ± 8.1 ****	**101.4 ± 12.7 ***	**113.9 ± 11.8 ****	**60.4 ± 7.9 ****
		O&O (249)	82.1 ± 14.5	112.2 ± 10.8	69.0 ± 8.2	178.4 ± 7.6	153.5 ± 17.4	56.1 ± 8.7	103.1 ± 12.7	120.4 ± 12.0	62.3 ± 7.7
	Females	NW (690)	87.9 ± 14.3	**102.3 ± 9.8 ***	**64.3 ± 7.8 ***	179.0 ± 8.0	**134.1 ± 14.6 ***	**54.6 ± 7.6 ***	103.0 ± 11.5	**108.5 ± 11.2 ***	59.9 ± 7.4
		O&O (68)	88.0 ± 15.0	106.5 ± 12.2	67.0 ± 8.4	179.3 ± 6.3	139.0 ± 15.9	56.7 ± 7.7	105.1 ± 12.5	112.3 ± 13.0	61.0 ± 8.7
16-18 years	Males	NW (1133)	**77.5 ± 13.0 ***	**110.1 ± 10.7 ****	68.5 ± 7.7	176.5 ± 6.1	**151.6 ± 18.1 ****	**55.5 ± 8.4 ***	**101.2 ± 12.5 ***	**118.1 ± 11.9 ****	**62.4 ± 8.2 ****
		O&O (137)	81.0 ± 14.5	114.1 ± 11.2	70.5 ± 8.2	176.5 ± 7.3	159.6 ± 20.9	57.2 ± 8.6	103.1 ± 11.2	123.3 ± 12.7	65.1 ± 7.9
	Females	NW (425)	86.2 ± 14.3	102.2 ± 10.5	**65.2 ± 7.7 ***	178.4 ± 6.3	**135.5 ± 15.5 ***	55.6 ± 5.6	107.9 ± 13.0	**108.8 ± 12.8 ***	60.2 ± 8.4
		O&O (26)	89.5 ± 16.8	105.4 ± 12.2	67.4 ± 6.3	177.0 ± 7.3	139.4 ± 15.9	55.6 ± 7.7	103.9 ± 13.1	112.7 ± 11.8	60.8 ± 7.6

Legend: HR: heart rate; BP: blood pressure; NW: normal weight; O&O: overweight and obese. * $p < 0.05$ between normal weight and overweight and obese subjects, ** $p < 0.001$ between normal weight and overweight and obese subjects. In parentheses, the number of subjects referred to the group.

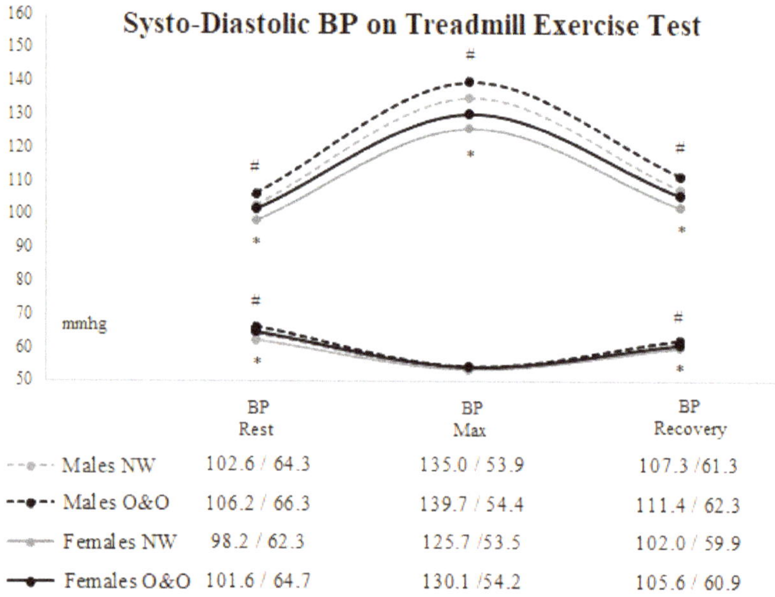

	BP Rest	BP Max	BP Recovery
Males NW	102.6 / 64.3	135.0 / 53.9	107.3 / 61.3
Males O&O	106.2 / 66.3	139.7 / 54.4	111.4 / 62.3
Females NW	98.2 / 62.3	125.7 / 53.5	102.0 / 59.9
Females O&O	101.6 / 64.7	130.1 / 54.2	105.6 / 60.9

Figure 4. Differences in blood pressure (BP) performances during treadmill GXT according to weight status and gender. Legend: BP: blood pressure; NW: normal weight; O&O: overweight and obese. * $p < 0.05$ between normal weight vs. overweight and obese females, # $p < 0.05$ between normal weight vs. overweight and obese males.

4. Discussion

The objective of this study was to describe the impact of excess weight on the cardiovascular parameters in young athletes; therefore, heart rate and blood pressure values resulting from a GXT were analyzed. Although the results of the present study were within normal range, they demonstrate how excess weight affects the cardiovascular parameters recorded both at rest and during physical exertion, even in young, healthy, and sporty people [22].

Epidemiologically, the prevalence of overweight and obesity is in agreement with previous studies carried out on young athletic subjects of the same territory [23–25].

Unlike previous studies [3], HR max was the only parameter not affected, both by analyzing the sample based on gender and the type of exercise performed. However, Italian guidelines declare that the achievement of 85% of age-predicted maximal HR is a main criterion to define a sufficient effort during GXT and generally, it is used in combination with other parameters to terminate the evaluation [20]. Although HR max shows no differences between the NW and O&O subjects, the magnitude of increase or decrease, studied with Δ HR max-rest and Δ HR max-rec, is influenced by the parameters of HR rest and HR recovery, both greater in subjects with excess weight [12]. In particular, the differences between NW and O&O in HR at rest start from the age group of 12–14 years in the male group. This confirms the results of previous studies conducted on non-sports subjects [26]; however, the HR rest of non-athletic obese children appears higher than the results obtained from O&O subjects of the present study conducted on a sample of young athletic subjects. The differences between NW and O&O in HR recovery also begin in the male group but already from 10–12 years. The comparison in HR recovery between the values of the sports subjects recorded after 4 min with those of the non-sports subjects recorded after 6 min from the end of the GXT describe a better performance in young athletes on average of 5 bpm for the subjects NW than for O&O [26]. Therefore, it is possible to speculate that sports practice can improve HR at rest and the recovery process after physical effort also in excess body weight condition. Furthermore, the differences in HR between NW and O&O can be shown early through physical exertion. In fact, the GXT results show these differences in a previous age group compared to those shown in resting conditions.

The parameter that appears most influenced by excess weight is the SBP, which undergoes an increase both in resting values, at the peak of the effort and after 4 min from the interruption of the GXT, probably due to a greater cardiac output in the O&O subjects [17]. As reported by previous studies [14], our results also show that older athletes have a lower SBP max value than younger ones. The physiological response of SBP during dynamic exercise usually involves an increase from 50 to 70 mmHg in young people whether they are normotensive or hypertensive [27]; however, Δ SBP max-rest shows data around 30–40 mmHg, with higher values in the exercise performed by the females and with cycle ergometer. This lower Δ SBP max-rest value obtained by young athletic subjects suggests a better relationship between cardiac output and the response of the vascular endothelium. The analysis of the Δ SBP max-rec results show less significant differences, therefore the SBP has slightly higher values but the effort in excess weight conditions has a reduced effect on the recovery of blood pressure in young athletes.

There is currently no behavior recognized as physiological in DBP's response to exercise; some published studies have shown a decrease in DBP, whereas others have reported no change in the levels observed at rest and a few have noted an increase in DBP [10,28]. Our DBP results shows a decrease behavior of about 10 mmHg upon reaching the peak of the exercise. Excess weight affects resting values in both sexes, while the values related to effort are mainly higher in males from 12 years old. In particular, the analysis of Δ DBP does not seem to add information in this context.

Body weight influences physical performance [26,29]; the results relating to the number of completed stages show lower values in O&O subjects. A condition that appears to reduce this behavior is the use of the cycle ergometer. This is probably because in young athletes, where there is no state of physical deconditioning, the support that the tool exerts on weight is somehow useful, avoiding any direct gravity action of the body on the ground.

The use of the cycle ergometer also promotes the reduction of the differences between NW and O&O in DBP values in females at the maximum effort and during recovery. However, it appears more effective in reducing chronotropic increase differences in males. No difference, in fact, was registered from the third stage, probably because of the body weight support action.

The present study shows strengths. First, the subjects were different from each other, they all belonged to the same center, and the evaluation methodology was standardized. The second is the homogeneity of the sample, all young people practiced organized sports at a competitive level.

The study has the limitation of being retrospective; this influences causality because the registered weight status is established in advance of the exercise test day. In addition, the study defines the weight status through the BMI: this method has the intrinsic limitation in distinguishing an overweight condition due to an excess of body fat from that due to a prevalence of muscle tissue. However, the authors chose this study design and the use of BMI in order to achieve this sample size with the homogeneity between the subjects described above.

This study did not distinguish children and adolescents by maturity stage; it could be inferred that the heterogeneity of the sample could alter the accuracy of the results. However, the objective of the study does not foresee the evaluation following growth phases, since this aspect involves an additional variable that would not allow the evaluation carried out on such a large sample [30].

5. Conclusions

Excess weight affects cardiovascular parameters both at rest and in response to physical exertion during youth. The cardiovascular parameters analyzed in the present study are heart rate and blood pressure: the effect that excess weight has on these parameters increases with the progress of growth.

The parameters that could help provide further information seem to be the differences (Δ) between rest and peak of effort and between peak of effort and after 4 min of recovery.

However, competitive sports seem to be able to keep these parameters within the normal range even in young overweight athletic subjects. This once again highlights the effectiveness of youth from a public health perspective.

Author Contributions: Conceptualization, G.M. and C.P.; methodology, G.M.; software, V.B.; formal analysis, V.B.; investigation, G.M., L.S., and L.T.; data curation, G.M.; writing—original draft preparation, G.M.; supervision, G.G.; project administration, P.C. All authors have read and agreed to the published version of the manuscript.

Funding: This research received no external funding.

Conflicts of Interest: The authors declare no conflicts of interest.

References

1. WHO. Obesity Fact Sheet. Available online: https://www.who.int/news-room/fact-sheets/detail/obesity-and-overweight (accessed on 5 March 2020).
2. Martini, G.; Riva, P.; Rabbia, F.; Molini, V.; Ferrero, G.B.; Cerutti, F.; Carra, R.; Veglio, F. Heart rate variability in childhood obesity. *Clin. Auton. Res.* **2001**, *11*, 87–91. [CrossRef] [PubMed]
3. Török, K.; Szelényi, Z.; Pórszász, J.; Molnár, D. Low physical performance in obese adolescent boys with metabolic syndrome. *Int. J. Obes. Relat. Metab. Disord.* **2001**, *25*, 966–970. [CrossRef] [PubMed]
4. Fletcher, G.F.; Ades, P.A.; Kligfield, P.; Arena, R.; Balady, G.J.; Bittner, V.; Coke, L.A.; Fleg, J.L.; Forman, D.E.; Gerber, T.C.; et al. Exercise standards for testing and training: A scientific statement from the American Heart Association. *Circulation* **2013**, *128*, 873–934. [CrossRef] [PubMed]
5. Thompson, P.D. Exercise prescription and proscription for patients with coronary artery disease. *Circulation* **2005**, *112*, 2354–2363. [CrossRef] [PubMed]
6. Pinkstaff, S.; Peberdy, M.A.; Kontos, M.C.; Finucane, S.; Arena, R. Quantifying exertion level during exercise stress testing using percentage of age-predicted maximal heart rate, rate pressure product, and perceived exertion. *Mayo Clin. Proc.* **2010**, *85*, 1095–1100. [CrossRef] [PubMed]
7. Bjelakovic, L.; Vukovic, V.; Jovic, M.; Bankovic, S.; Kostic, T.; Radovanovic, D.; Pantelić, S.; Zivkovic, M.; Stojanovic, S.; Bjelakovic, B. Heart rate recovery time in metabolically healthy and metabolically unhealthy obese children. *Phys. Sportsmed.* **2017**, *45*, 438–442. [CrossRef]
8. McArdle, W.D.; Katch, F.I.; Katch, V.L. *Exercise Physiology: Nutrition, Energy, and Human Performance*; Lippincott Williams & Wilkins: Baltimore, MD, USA, 2014.
9. Syme, A.N.; Blanchard, B.E.; Guidry, M.A.; Taylor, A.W.; Vanheest, J.L.; Hasson, S.; Thompson, P.D.; Pescatello, L.S. Peak systolic blood pressure on a graded maximal exercise test and the blood pressure response to an acute bout of submaximal exercise. *Am. J. Cardiol.* **2006**, *98*, 938–943. [CrossRef]
10. Paridon, S.M.; Alpert, B.S.; Boas, S.R.; Cabrera, M.E.; Caldarera, L.L.; Daniels, S.R.; Kimball, T.R.; Knilans, T.K.; Nixon, P.A.; Rhodes, J.; et al. Clinical stress testing in the pediatric age group: A statement from the American Heart Association Council on Cardiovascular Disease in the Young, Committee on Atherosclerosis, Hypertension, and Obesity in Youth. *Circulation* **2006**, *113*, 1905–1920. [CrossRef]
11. Cooper, D.M.; Weiler-Ravell, D.; Whipp, B.J.; Wasserman, K. Growth-related changes in oxygen uptake and heart rate during progressive exercise in children. *Pediatr. Res.* **1984**, *18*, 845–851. [CrossRef]
12. Singh, T.P.; Rhodes, J.; Gauvreau, K. Determinants of heart rate recovery following exercise in children. *Med. Sci. Sports Exerc.* **2008**, *40*, 601–605. [CrossRef]
13. Easley, E.A.; Black, W.S.; Bailey, A.L.; Lennie, T.A.; Sims, W.J.; Clasey, J.L. Recovery Responses to Maximal Exercise in Healthy-Weight Children and Children with Obesity. *Res. Q. Exerc. Sport* **2018**, *89*, 38–46. [CrossRef] [PubMed]
14. Szmigielska, K.; Szmigielska-Kapłon, A.; Jegier, A. Blood pressure response to exercise in young athletes aged 10 to 18 years. *Appl. Physiol. Nutr. Metab.* **2016**, *41*, 41–48. [CrossRef] [PubMed]
15. Lintu, N.; Tompuri, T.; Viitasalo, A.; Soininen, S.; Laitinen, T.; Savonen, K.; Lindi, V.; Lakka, T.A. Cardiovascular fitness and haemodynamic responses to maximal cycle ergometer exercise test in children 6–8 years of age. *J. Sports Sci.* **2014**, *32*, 652–659. [CrossRef]

16. Lintu, N.; Viitasalo, A.; Tompuri, T.; Veijalainen, A.; Hakulinen, M.; Laitinen, T.; Savonen, K.; Lakka, T.A. Cardiorespiratory fitness, respiratory function and hemodynamic responses to maximal cycle ergometer exercise test in girls and boys aged 9–11 years: The PANIC Study. *Eur. J. Appl. Physiol.* **2015**, *115*, 235–243. [CrossRef] [PubMed]

17. de Sousa, G.; Hussein, A.; Trowitzsch, E.; Andler, W.; Reinehr, T. Hemodynamic responses to exercise in obese children and adolescents before and after overweight reduction. *Klin. Padiatr.* **2009**, *221*, 237–240. [CrossRef] [PubMed]

18. Barker, A.R.; Armstrong, N. Exercise testing elite young athletes. *Med. Sport Sci.* **2011**, *56*, 106–125. [CrossRef] [PubMed]

19. Mascherini, G.; Galanti, G.; Massetti, L.; Calà, P.; Modesti, P.A. Growth Charts for Height, Weight, and BMI (6–18 y) for the Tuscany Youth Sports Population. *Int. J. Environ. Res. Public Health* **2019**, *16*, 4975. [CrossRef]

20. Biffi, A.; Delise, P.; Zeppilli, P.; Giada, F.; Pelliccia, A.; Penco, M.; Casasco, M.; Colonna, P.; D'Andrea, A.; D'Andrea, L.; et al. Italian cardiological guidelines for sports eligibility in athletes with heart disease: Part 1. *J. Cardiovasc. Med.* **2013**, *14*, 477–499. [CrossRef]

21. Cole, T.J.; Lobstein, T. Extended international (IOTF) body mass index cut-offs for thinness, overweight and obesity. *Pediatr. Obes.* **2012**, *7*, 284–294. [CrossRef]

22. Coledam, D.H.C.; Ferraiol, P.F.; Pires, R.; Júnior Greca, J.P.A.; Oliveira, A.R. Overweight and obesity are not associated to high blood pressure in young people sport practitioners. O sobrepeso e a obesidade não estão associados com a pressão arterial elevada em jovens praticantes de esportes. *Ciência Saúde Coletiva* **2017**, *22*, 4051–4060. [CrossRef]

23. Mascherini, G.; Petri, C.; Ermini, E.; Bini, V.; Calà, P.; Galanti, G.; Modesti, P.A. Overweight in Young Athletes: New Predictive Model of Overfat Condition. *Int. J. Environ. Res. Public Health* **2019**, *16*, 5128. [CrossRef] [PubMed]

24. Petri, C.; Mascherini, G.; Bini, V.; Toncelli, L.; Armentano, N.; Calà, P.; Galanti, G. Evaluation of physical activity and dietary behaviors in young athletes: A pilot study. *Minerva Pediatr.* **2017**, *69*, 463–469. [CrossRef] [PubMed]

25. Petri, C.; Mascherini, G.; Bini, V.; Anania, G.; CAL, P.; Toncelli, L.; Galanti, G. Integrated total body composition versus Body Mass Index in young athletes. *Minerva Pediatr.* **2016**. Available online: https://www.minervamedica.it/it/riviste/minerva-pediatrica/articolo.php?cod=R15Y9999N00A16040801 (accessed on 5 March 2020).

26. Norman, A.C.; Drinkard, B.; McDuffie, J.R.; Ghorbani, S.; Yanoff, L.B.; Yanovski, J.A. Influence of excess adiposity on exercise fitness and performance in overweight children and adolescents. *Pediatrics* **2005**, *115*, e690–e696. [CrossRef]

27. Pate, R.R.; Matthews, C.; Alpert, B.S.; Strong, W.B.; DuRant, R.H. Systolic blood pressure response to exercise in black and white preadolescent and early adolescent boys. *Arch. Pediatr. Adolesc. Med.* **1994**, *148*, 1027–1031. [CrossRef]

28. Ahmad, F.; Kavey, R.E.; Kveselis, D.A.; Gaum, W.E.; Smith, F.C. Responses of non-obese white children to treadmill exercise. *J. Pediatr.* **2001**, *139*, 284–290. [CrossRef]

29. McGavock, J.M.; Torrance, B.D.; McGuire, K.A.; Wozny, P.D.; Lewanczuk, R.Z. Cardiorespiratory fitness and the risk of overweight in youth: The Healthy Hearts Longitudinal Study of Cardiometabolic Health. *Obesity* **2009**, *17*, 1802–1807. [CrossRef]

30. Campa, F.; Silva, A.M.; Iannuzzi, V.; Mascherini, G.; Benedetti, L.; Toselli, S. The Role of Somatic Maturation on Bioimpedance Patterns and Body Composition in Male Elite Youth Soccer Players. *Int. J. Environ. Res. Public Health* **2019**, *16*, 4711. [CrossRef]

Article

Mindfulness and Self-Regulation Strategies Predict Performance of Romanian Handball Players

Daniela Popa [1,*], Veronica Mîndrescu [2,*], Teodora-Mihaela Iconomescu [3,*] and Laurentiu-Gabriel Talaghir [4,5,*]

[1] Psychology, Education and Teacher Training Department, Transilvania University of Brasov, Bulevardul Eroilor 29, 500036 Brașov, Romania

[2] Motor Performance Department, Transilvania University of Brasov, Bulevardul Eroilor 29, 500036 Brașov, Romania

[3] Sports Games and Physical Education Department, Dunarea de Jos University of Galati, Domneasca street no. 47, 800008 Galați, Romania

[4] Individual Sports and Physical Therapy Department, Dunarea de Jos University of Galati, Domneasca street no. 47, 800008 Galați, Romania

[5] South Ural State University, Prospekt Lenina no. 76, 454080 Chelyabinsk Oblast, Russia

* Correspondence: danapopa@unitbv.ro (D.P.); mindrescu.veronica@unitbv.ro (V.M.); ticonomescu@ugal.ro (T.-M.I.); gtalaghir@ugal.ro (L.-G.T.)

Received: 6 April 2020; Accepted: 28 April 2020; Published: 1 May 2020

Abstract: Previous studies on handball players' performance are focused more on influence of physical, physiological factors and tactical strategies and less on the influence of cognitive, metacognitive and attentional regulation strategies. Performance can be achieved by attentional and emotional regulation alongside cognitive, metacognitive and procedural regulation strategies. This study explores the association between self-regulation strategies, mindfulness practice and performance. The sample consists of 288 Romanian handball players. The participants were 30% male and 70% female, with age between 12.01 and 14 years old, divided in three categories. The quantitative research design is descriptive and transversal. The method was survey based on questionnaires. There were interesting results found in different age categories and different performance levels. The variables (state mindfulness of body, self-monitoring, and self-efficacy) explained 87% of the variance in sports performance, in a hierarchical multiple regression. The research findings indicated that handball players with a high level of acceptance of one's own thoughts and emotions, non-judging present-moment awareness, conscious monitoring the execution of movements, and confidence in their abilities to succeed could have more chances to achieve the desired performance.

Keywords: mindfulness in physical activity; performance; self-efficacy; self-monitoring

---

1. Introduction

The practice of mindfulness (conscious attention of the present) influences the modalities of using self-regulation strategies, giving them a new level of efficiency, with the two behaviours having a strong impact on performance in general sports [1]). For example, Noetel and his team highlighted that mindfulness can be a path to enhance athletic performance by optimally adjusting attentional processes, emotions and cognitions [2].

Furthermore, in a meta-analytical review of nine trials, with 290 healthy sportive participants, Bühlmayer et al. [3] found that mindfulness practice positively influences the athletes' performance outcomes. Self-regulation strategies have been also linked to increases in performance [4]. Athletes with a high level of self-regulation manifest a greater degree of awareness of their strengths and weaknesses and make a greater effort to reach their goals [5].

There are researches that separately investigate the influence of mindfulness [2,6,7] and self-regulation [8,9] on performance. However, the researchers did not stop here; they wanted to find out if these two variables together influence performance. There are studies that have shown the relations between mindfulness, negative self-monitoring and emotion regulation [10], and the fact that mindfulness supports efficient emotion regulation, attention regulation and executive functioning [11]. It seems that mindfulness and self-regulation are interrelated components that influence performance, especially in team sports as they regulate social dynamics [12,13].

However, while previous studies only focused on these factors separately, we argue that in order to fully understand the way in which these psychological aspects influence the sports performance we need to consider both mindfulness and self-regulation strategies, at the same time.

The concept of self-regulation makes reference to "the thoughts, feelings, and self-generated actions that are systematically oriented towards reaching goals" [14]: 215. Self-regulation of learning implies a cyclical process, in which a person plans a task, monitors performance, and reflects upon the result. The cycle repeats when the person uses reflection for adapting and preparing for the next task [15]. In the activity planning stage, a sportsman establishes not only the goals he wishes to reach, but also a work strategy. These behaviours are not enough for a person to reach success; frequent monitoring of the activity's development, evaluation of each component and comparison with an initial target, and redefining a work strategy, if necessary, are needed. Thus, self-regulation supposes focusing one's attention on the goal proposed, making constant and focused efforts, which make self-regulation strategies draw significantly closer to the practice of mindfulness.

Recent studies have proved the efficiency of using an instructional design, typical for self-regulated learning, in teaching sports abilities in physical education, highlighting the important role of imitative practice [16]. Also, pursuing multiple goals and self-recording in the context of physical education improves performance more than in the situation in which only one is practised [17]. The conclusion of a recent literature review regarding the applicability of self-regulation in sports contexts highlighted that self-regulation can be taught, that highly performing sportsmen are highly self-regulated, and the necessity of a relation between self-regulation and other social and individual factors [18]. One of these individual factors can be the practice of mindfulness.

Also, research in this field has focused on categories of sportsmen, such as: athletes, soccer players, gymnasts, tennis players, volleyball players, and pupils who study physical education, but too little or not at all on those who play handball.

Mindfulness is defined as "the awareness that emerges through paying attention on purpose, in the present moment, and non-judgmentally to the unfolding of experience moment by moment" [19], that is, an acceptance of current experience with curiosity and openness [20]. Often, the concept of mindfulness is considered a form of emotional and attention regulation. Both emotions and attention energetically support the activity of an individual.

Recent scientific approaches to mindfulness have focused on investigating the effects that mindfulness practice can have both in different, clinical contexts [21], school contexts [22], and on the types of effects, such as the reduction of negative affects [23], the reduction of stress, the improvement of a good psychological state [24], sport performance enhancement [25], complex skill acquisition and sustainability, positive work-related outcomes, work-family balance, and job satisfaction [26].

Mindfulness practice, focused on the methods of maintaining awareness through intention, attention, and attitude, is different from dispositional mindfulness, which represents the tendency of currently being mindful [27]. Mindfulness practice regulates attention, affect, and behaviour, thus facilitating obtaining a high level of perseverance and determination.

Dispositional mindfulness is investigated in relation with sport-specific coping skills, proving that it has an important role in reducing rumination and other negative emotions, helping athletes to be aware of and to understand the inhibiting potential of negative thoughts and emotions [28].

Recent scientific results show that the combination of a sport with mindfulness practice allows for obtaining some remarkable improvements in the level of self-esteem, resilience, and happiness of teenagers [29]. The conclusion of a recent systematic review shows that an impressive number of studies have reported effects of practising mindfulness on the improvement of the flux, performance, and diminishing of anxiety before a contest. However, the same study signalled the necessity of some studies with an internal validity less limited than those studied [2]. Exercises in mindfulness practice allow an individual to explore movements, objects, and situations with undivided attention, and to act with awareness, without making value judgments that can embezzle a mind-set and diminish the level of performance.

Handball is a complex team sport in which individual performances are used as well as the interactions between teammates, tactics and strategies [30]. Previous studies on handball players' performance are focused more on the influence of the physical [31,32], physiological factors [33] and tactical strategies [34] and less on the influence of cognitive, metacognitive, and attentional regulation strategies.

Sports performance is in itself only a result, a score, or a final product [35]. However, to reach these goals, the development of a process of intense and long preparation is necessary. This process depends on subjective, socio-cultural, or environmental factors. Subjective factors are often divided into biological and psychological factors, although they often condition one another. Psychological factors have long been studied, both individually and in an interconnected manner, but the combination of factors and the extent of their influence becomes a personal equation of each individual.

Self-regulation of behaviour, a current practice of mindfulness methods, facilitates maintaining and enhancing a state of well-being and improves the focus of attention and speed of reaction, and allows an increase in the number of objects and persons present in the field of attention. All these conditions facilitate improving performance.

Careful and constant planning of activity, so as to allocate enough resources and time to practising the abilities that we wish to form or develop "is a mindful way to approach a new learning task, but students do not always mindfully regulate" their activities [36]: 425).

Small personal habits may turn either into the allies of a sportsman or obstacles in acquiring performance. A careful monitoring of activity, a level of self-efficacy appropriate for personal skills and abilities, as well as a mindful attitude allow the setting of a sportsman towards success and performance.

This study contributes to the existing literature by addressing the limitations of previous research, focusing on investigating strategies of regulating behaviour, which are important in sport, especially in handball. These are the self-regulation of emotions, attention, thoughts, attitudes (mindfulness), planning activity, reflection, self-monitoring, effort, self-evaluation, and self-efficacy level. We expand knowledge on factors that influence sports performance by simultaneously investigating some of the important psychological aspects.

Our primary goal is to explore the association between self-regulation, mindfulness strategies, and performance in Romanian handball players. The secondary objectives of this study are to verify if gender and the three age categories lead to different performance, different levels of self-regulation, and different practices of mindfulness.

The main hypothesis is that mindfulness and self regulation strategies play a significant role in predicting handball performance.

2. Materials and Methods

2.1. Participants

The sample consisted of 288 participants. Of the total number of 352 club members, 288 participants were recruited (81.82%). The other 64 club members (18.18%) who were not included in the study were comprised of the following categories: 54 members (15.34%) aged between 7 and 9 years (being in the initiation phase of performance handball), 4 members (1.14%) not present during the study period (due to the disease), and 6 members (1.7%) who did not complete for personal reasons. The gender distribution of the sample was 30% male and 70% were female.

The participants were divided in three age categories, the first comprising 83 sportsmen with ages between 10 and 12 years old, representing 28.8% of the total. The second age category, between 12.01 and 14 years old, was made up of 109 participants (37.8% of the total), and the third category, between 14.01 and 16 years old comprised 96 handball players, representing 33.3% (overall mean age was 13 years, the SD was 1.79, the minimum age was 10, and the maximum age was 16). Members of the junior teams of two clubs in Romania were recruited for this study.

The committees running the two clubs and the coaches involved approved this study. The parents of the participants filled in informed consent forms. Each participant gave his or her informed consent before starting the study. The participation was voluntary and research did not include any risk for the participants. The procedure of applying the instruments was the following: One of the authors of the article presented the aim of the research to the participants, its implications, the assignments of the participants, and their rights as participants. Those who wished to participate filled in the informed consent forms and filled in the instruments of the research. Junior handball players who did not wish to participate in the research left the room, because the moment of applying was at the end of training.

Four instruments were used: a questionnaire with questions referring to demographic characteristics (age, gender, type of sport practised, and the number of years of experience in this sport), data related to the level of performance attained and the number of weekly trainings, social contextual data; Self-Regulation of Learning Self-Report Scale (SRL-SRS; [37]) and The State Mindfulness Scale for Physical Activity (SMS-PA; [38]), and an assessment/hierarchy of the level of performance made by the coach. The time for filling in the instruments was 15–20 min at the most.

2.2. Measures

Results in Competitions

Respondents mentioned the level of ranking attained in competitions over the last year. They mentioned the level of competitions in which they participated: local, regional, national, international, their role in the team, if they were active or reserves, and the number of injuries.

2.3. Level of Performance Through the Hierarchy Made by the Coach

Each coach made a range of performances attained by participants on a five-level scale: very good, good, medium, weak, and very weak. For obtaining this indicator, the coach evaluated the body details and motor qualities (the size of the palm, reaction speed, movement speed, skills level, force, and resistance to effort).

2.4. Performance Level

The variable that expresses the level of performance attained by a participant represents a composite score made up of the arithmetic mean of the scores obtained for the variables: the level of ranking attained in competition over the last year and the level of performance on the scale made by the coach.

For Juniors IV (10–12 years), in measuring performance, the dribbling trial through flagpoles was used. This was done as follows: on the length of the handball court 7 flagpoles were set in a straight line, the first flagpole at a distance of 6 m from the start line and the last flagpole at a distance of 24 m before the finish line; all the other 5 flagpoles had a 3 m distance between them. The athletes ran the 30 m distance in a multiple dribbling, with the ball in permanent control. For the girls, the following performances were taken into account: maximum (1) +/−9.7 s, intermediary (2) +/−10 s, and minimum (3) +/−10.3 s. For the boys the performances were: maximum (1) +/−9 s, intermediary (2) +/−9.3 s, and minimum (3) +/−9.6 s.

For Juniors III (12–14 years), in measuring performance, the dribbling trial while running in a triangle was used. The triangle was marked as follows: the basis of the triangle was a straight line of 3 m of a semicircle of 6 m; on this basis one draws from its middle a vertical line of 3 m which led to the pointed 9 m semicircle and represented the height of the triangle. The sides of the triangle were drawn by uniting the three obtained points. In the tip of the triangle and tangential with it a circle of 30 cm in diameter was marked. Initially the athlete faced the tip of the triangle having the left foot within the left circle from the basis of the triangle. The athlete ran the distance step by step to the right until he/she touched the other circle with his/her foot from the tip of the triangle; then, again, step by step, moving forward into the circle from the tip of the triangle, which was compulsory to touch with any foot, after which, again, step by step, moving backwards until he/she reached with one foot the circle from which he/she originally started. This was done in two complete laps. For the girls the following performances were taken into account: maximum (1) +/−16.6 s, intermediary (2) +/−16.9 s and minimum (3) +/−17.2 s. For the boys the performances were: maximum (1) +/−16 s, intermediary (2) +/−16.3 s and minimum (3) +/−16.6 s.

2.5. Social Contextual Sports Characteristics

The participants indicated changes of club, team, or/and coach over the last year and the number of weekly hours allotted to training. We decided to take into account these contextual characteristics too, as they moderately affect the performance of a sportsman [37].

2.6. Self-Regulation of Learning Self-Report Scale

The SRL-SRS Scale [37] was translated into Romanian following the back-translation procedure [39] and culturally adapted, using synonymous terms closer to the common language of Romanian youth.

The questionnaire developed by Bartulovic, Young, & Baker [37] contains 48 items of the 50 original SRL-SRS Scale [40]. The items were supplemented with a context-specific approach and are placed on seven-point scales (1—almost never, 4—sometimes, and 7—almost always). The six subscales of Toering et al.'s Scale [40] indicate Cronbach's α coefficients as: "planning = 0.81, self-monitoring = 0.73, evaluation = 0.82, reflection = 0.78, effort = 0.85, and self-efficacy = 0.81" ([41]: 212). Bartulovic, Young, & Baker's model [37] demonstrates a good fit χ^2 (16, N = 268) = 444.0, $p \leq 0.001$) and the subscales were consistent with those reported by Toering et al. [40].

2.7. The State Mindfulness Scale for Physical Activity

The SMS-PA is a 12-item questionnaire, assessing two dimensions: state mindfulness of mind (6 items) and state mindfulness of body (6 items). The scale demonstrated a strong internal consistency reliability ($\alpha > 0.80$) and the model revealed a good fit χ^2 (53) = 176.22, $p < 0.01$, SRMR −0.06, CFI = 0.96, TLI = 0.95, WRMR = 1.42).

The items were placed on a 5-pt scales (1—not at all, 3—moderately and 5—very much).

2.8. Statistical Analyses

All statistical analyses were calculated using IBM SPSS Statistics 21. Correlational analyses were carried out and non-parametric tests were used for comparing the subgroups of sample. Finally, a linear regression analysis was conducted in order to discover which of the variables explained the variance of the sports performance of respondents.

3. Results

Descriptive Statistics

A large portion of the participants reported a fairly high level of the values for all analysed variables, high scores prevailing, which was to be expected given that respondents were performance sportsmen (Table 1. Characteristics of the sample study).

Table 1. Characteristics of the sample study (N = 288).

Characteristics	Mean (SD)	Minimum	Maximum	Median	Skewness	Kurtosis
Performance	2.75 (0.48)	1	3	3.00	−1.71	2.09
Planning	50.66 (8.82)	25	63	52.00	−0.67	0.26
Self-monitoring	46.86 (7.11)	20	56	48.00	−1.09	1.18
Evaluation	40.78 (6.90)	18	49	43.00	−0.99	0.70
Reflection	30.46 (4.38)	15	35	31.00	−1.31	1.56
Effort	63.48 (7.43)	30	70	66.00	−1.75	3.48
Self-efficacy	55.11 (6.22)	31	63	56.00	−1.09	1.50
State mindfulness of mind	17.76 (4.19)	4	24	19.00	−1.21	0.97
State mindfulness of body	19.45 (4.26)	6	24	21.00	−1.30	0.84

Std. Error of Skewness 0.144; Std. Error of Kurtosis 0.28.

In order to check the existence of significant correlations among the studied variables, the Spearman rho correlation coefficient was calculated, as the variables were not normally distributed. The condition of normality of the distribution of variables was checked by means of the Kolmogorov–Smirnov Z test. The condition of the linearity of the relations among the variables was checked by inspecting the cloud of points.

Spearman's Rho correlations coefficient between research variables, according to the results obtained, found positive correlations between the level of performance, state mindfulness of mind, state mindfulness of body, planning, self-monitoring, evaluation, reflection, effort, and self-efficacy level (Table 2). Thus, we can state that the greater the level of performance attained, the greater the concern for mindfulness and the more strategies of self-regulating sports activity the respondent possessed. We discovered weak negative correlations between the variable age and the level of performance, state mindfulness of mind, and state mindfulness of body. Thus, sportsmen with lower ages attain high performances more easily in this sport and practise mindfulness more than the older ones.

Table 2. Spearman's Rho correlations between research variables.

	1	2	3	4	5	6	7	8	9	10
1. Performance	-									
2. State Mindfulness of Mind	0.735 **	-								
3. State Mindfulness of Body	0.736 **	0.976 **	-							
4. Planning	0.589 **	0.417 **	0.436 **	-						
5. Self-Monitoring	0.732 **	0.534 **	0.555 **	0.756 **	-					
6. Evaluation	0.581 **	0.468 **	0.482 **	0.711 **	0.741 **	-				
7. Reflection	0.449 **	0.392 **	0.399 **	0.613 **	0.670 **	0.788 **	-			
8. Effort	0.399 **	0.301 **	0.310 **	0.558 **	0.614 **	0.724 **	0.779 **	-		
9. Self-Efficacy	0.435 **	0.366 **	0.389 **	0.616 **	0.696 **	0.659 **	0.666 **	0.713 **	-	
10. Age	−0.127 *	−0.140 *	−0.156 **	0.050	−0.032	−0.017	−0.077	−0.104	0.030	-

** Correlation is significant at the 0.01 level (2-tailed), * Correlation is significant at the 0.05 level (2-tailed), N = 288, Age = Age categories.

Although this statement may seem risky, it may have some explanations. On the one hand, the 10–12 age group has certain personality traits that allow them to better cooperate in team sports. Developmental psychology considers that at this stage of life, peers and teammates are the main reference group [42]. As athletes grow, the internal frame of reference becomes much more important, and the focus of attention is shifted from colleagues to themselves, the individual striving to perform better individually so that she/he is satisfied. Also, the main concerns change, the older ones being less interested in the team and more interested in the dyads. Additionally, mindfulness and self-regulation regulate social dynamics. This has direct consequences on performance, especially in team sports such as handball [13]. However, the correlation coefficient was extremely small, so no strong statement may be made.

The data were examined to verify if the results of sportsmen were influenced by gender variable; we applied the Mann-Whitney U test for two independent samples.

As noted in Table 3, the results showed that there were no significant differences among the groups of participants, thus we consider that the variable gender has no effect on performance, self-regulation strategies used, or mindfulness practice.

Table 3. Mann-Whitney U test statistics, grouping variable gender.

	Performance	SMM	SMB	Planning	SM	Evaluation	Reflection	Effort	SE
Mann-Whitney U	4290.50	4494.50	4495.50	4341.50	4240.50	4318.500	4246.50	4284.50	4003.50
Wilcoxon W	10285.50	10489.50	7981.50	10336.50	7726.50	7804.50	7732.50	10279.50	7489.50
Z	−0.88	−0.07	−0.07	−0.47	−0.74	−0.53	−0.73	−0.63	−1.36
Asymp. Sig. (2-tailed)	0.37	0.93	0.94	0.63	0.45	0.59	0.46	0.52	0.17

Grouping Variable: Gender. SMM = state mindfulness of mind, SMB = state mindfulness of body, SM = self-monitoring, SE = self-efficacy.

Another aim of the research was to inspect if the three age categories lead to different performance, different levels of self-regulation, and different practices of mindfulness. The Kruskal-Wallis H test was applied, the non-parametric equivalent of the ANOVA one-way test. This test was applied as the variables did not follow normal distributions.

There are significant differences between the categories of participants depending on age, for the variables state mindfulness of mind and state mindfulness of body (Table 4). However, this test did not show which category of age practises mindfulness more intensely, therefore we applied the Mann-Whitney U test in order to discover among which of the three age categories the differences were significant, comparing them two by two, adjusting the significance threshold depending on the number of comparisons through the Bonferroni method (three in this case), so that $\alpha = 0.05/3 = 0.01$.

Table 4. Kruskal-Wallis test statistics, grouping variable age category.

K-W Test [a]	Performance	SMM	SMB	Planning	SM	Evaluation	Reflection	Effort	SE
Chi^2	4.71	7.59	9.94	1.90	1.65	0.60	4.10	3.20	2.04
df	2	2	2	2	2	2	2	2	2
A. Sig.	0.09	0.02	0.00	0.38	0.43	0.73	0.12	0.20	0.35

[a] Grouping Variable: Age category. SMM = state mindfulness of mind, SMB = state mindfulness of body, SM = self-monitoring, SE = self-efficacy, Chi^2 = Chi-square, A. Sig. = asymptomatic significance level.

There were no significant differences among the age groups 10–12 years old and 12–14 years old regarding the practice of mindfulness of body (U = 4495.50, z = −0.07, p = 0.94, r = 1.75, Mdn_{10-12} = 21.00, Mdn_{12-14} = 22.00) and state mindfulness of mind (U = 4494.50, z = −0.07, p = 0.93, r = 1.75, Mdn_{10-12} = 19.00, Mdn_{12-14} = 20.00).

There were significant differences among age groups 10–12 years old and 14–16 years old regarding the mindfulness of body practice (U = 3154.00, z = −2.42, p = 0.01, r = 0.17, Mdn_{10-12} = 21.00, Mdn_{14-16} = 20.00).

Further, there were significant differences between the age groups 12–14 years old and 14–16 years old regarding the mindfulness of body practice (U = 3980.00, z = −2.97, p = 0.00, r = 0.20, Mdn_{12-14} = 22.00, Mdn_{14-16} = 20.00) and state mindfulness of mind (U = 4139,00, z = −2.59, p = 0,00, r = 0.17, Mdn_{12-14} = 20.00, Mdn_{14-16} = 18.00). For these two variables, the age category 14–16 years old obtained the lowest scores.

As the number of sportsmen with low performance was poorly represented, we checked the existence of significant differences among sportsmen with average performance and those with high performance, regarding the studied variables.

The results obtained following the application of the Mann-Whitney U test (Table 5) show that there were significant differences between the two sub-groups.

Table 5. Mann-Whitney U test statistics, grouping variable performance.

K-W Test [a]	SMM	SMB	Planning	SM	Evaluation	Reflection	Effort	SE	SRL
Mann-Whitney U	0.00	0.00	1438.00	0.00	1544.00	2904.00	3250.00	2822.00	574.00
Wilcoxon W	1830.00	1830.00	3268.00	1830.00	3374.00	4734.00	5080.00	4652.00	2404.00
Z	−11.96	−11.98	−9.32	−11.91	−9.15	−6.74	−6.13	−6.86	−10.86
Asymp. Sig. (2-tailed)	0.00	0.00	0.00	0.00	0.00	0.00	0.00	0.00	0.00

[a]. Grouping Variable: Performance.

With the focus to explore the efficiency of the new explanatory model of sports performance, on the basis of the variables mindfulness and self-regulation, we applied the method of hierarchical multiple regression.

The results obtained show that model three, which contains the variables state mindfulness of body, self-monitoring, and self-efficacy, explained sports performance best, the contribution of the three variables being significant. In the case of model two, $R^2_{adjusted} = 0.87$, which means it explained 87% of the variance in sports performance, the global effect being of a high level (Table 6). Also, state mindfulness of body had the highest explanatory weight of the three variables of the model, followed by the two self-regulation components mentioned above.

Table 6. Regression model summary, dependent variable: performance.

Predictor	b	b 95% CI [LL, UL]	beta	beta 95% CI [LL, UL]	Sr^2	Sr^2 95% CI [LL, UL]	r	Fit
(Intercept)	0.53 **	[0.36, 0.71]						
Mindfulness	0.07 **	[0.06, 0.08]	0.64	[0.57, 0.71]	0.16	[0.11, 0.20]	0.91 **	
Self monitoring	0.03 **	[0.02, 0.03]	0.41	[0.33, 0.49]	0.05	[0.03, 0.07]	0.84 **	
Self efficacy	−0.01 **	[−0.01, −0.00]	−0.11	[−0.17, −0.06]	0.01	[0.00, 0.01]	0.43 **	
								$R^2 = 0.879$ **
								95% CI[NA,NA]

A significant b-weight indicates the beta-weight and semi-partial correlation were also significant. b represents unstandardized regression weights. *beta* indicates the standardized regression weights. sr^2 represents the semi-partial correlation squared. r represents the zero-order correlation. *LL* and *UL* indicate the lower and upper limits of a confidence interval, respectively. * indicates $p < 0.05$. ** indicates $p < 0.01$.

4. Discussion

The overarching aim of this research was to explore the association between self-regulation, mindfulness strategies and performance in a group of Romanian handball players. This study responds to the need noted in the literature to investigate the cognitive and social factors that may influence handball teams [30]. As previous researches showed [9,11], it seems that mindfulness and self-regulation are interrelated and predict performance.

The association between the two aspects, the emotional and attentional regulation (mindfulness) and the self-regulation of the behaviour in the sport activity of the handball players has been barely studied [43]. By further comparing the results obtained for different levels of performance, operational differences in these psychological factors can be highlighted.

One of the most interesting ideas that emerges from the analysis of the obtained results is that a very high level of self-efficacy and confidence in one's own powers diminishes performance. This point of view is consistent with the results of recent studies ([44,45], which show that self-efficacy negatively influences performance following the exertion of cognitive control.

The very high level of self-efficacy can be associated with the risk of formation of slightly erroneous automatisms too early, diminishing the personal involvement in the task preparation process.

5. Gender, Self-Regulation and Mindfulness

The results obtained show that the gender variable does not influence the answers of sportsmen to any of the dimensions considered. There are no statistically significant differences among the categories of respondents depending on the gender variable. In recent studies that examined gender differences in mindfulness intervention on respondents of the same age as the population of this study, the researchers discovered the existence of gender differences [46]. Also, since gender differences in self-regulation have been reported [47,48], we verified the existence of these relationships in the present study. The lack of such relationships may be due to the characteristics of the small sample.

6. Age Differences

Conversely, age influences the level of performance, state mindfulness of mind and state mindfulness of body. Contrary to expectations, the younger the sportsmen, the better the sports result they obtain. Also, they practise mindfulness techniques more often and more intensely, or perhaps they present a higher level of openness and acceptance of daily experiences. Another possible explanation can be that at the early age of sportsmen behavioural automatisms were not formed, so it is easier to consciously monitor movements and cognitions.

7. Self-Regulation and Performance

Although the planning variable correlates strongly and statistically significant with the performance variable ($\rho = 0.589$, $p < 0.01$), this does not become a criterion variable in the regression equation of the performance variable. The same statement can be made regarding the effort ($\rho = 0.399$, $p < 0.01$) and reflection variables ($\rho = 0.449$, $p < 0.01$).

Conversely, self-monitoring and self-efficacy are part of the predictors of sport performance. The results obtained show that planning the activity, reflection on the game, and the quantity of effort made are necessary but not sufficient to explain success. The results reinforce the theory of reinvestment perspective [49]. Overthinking and the tendency to control every step and move is detrimental to performance, especially in stressful contexts.

In the case of this group of handball players, careful self-monitoring of cognitive, meta-cognitive strategies is more important, accepting them calmly and consciously, focusing attention on details typical to handball games. Thus, a person has the opportunity to reduce the quantity of automated behaviours by focusing on exercising skills and reducing behavioural habits that are less efficient or appropriate.

A consistent literature emphasized that self-efficacy is a significant predictor of sport performance [50,51].

A high level of trust in oneself allows addressing challenges that suppose assuming risks and surpassing the level of performance at a certain moment.

This finding is consistent with earlier researches with volleyball players [52] showing that self-efficacy supposes the belief of a person in the power of his own abilities to execute a certain action. This allows a sportsman to have the correct interpretation of the quantity of effort that has to be made to reach the objectives desired, its dosing over time, and using coping mechanisms depending on the situation.

8. Mindfulness and Performance

Mindfulness of body measures the range to which an individual is fully aware of his physical perceptions, endurance, necessary efforts, muscular engagement, or movements of the body [53]. This concept highlights the deep connection with the physical body, developing insight in a balanced, non-judgmental, curious, and accepting manner, preventing relapses [54].

Mindfulness of mind focuses on being conscious of emotions and thought patterns, and on following the flux of thoughts, irrespective of their emotional quality. In this study, we correlate self-monitoring ($\rho = 0.534$, $p < 0.01$) and evaluation ($\rho = 0.468$, $p < 0.01$), with planning ($\rho = 0.417$, $p < 0.01$).

Limitations

Some of the limits of the research are related to self–reported data for understanding the concepts of mindfulness and self-regulation, the convenience sample typically biases [55,56]. Other limitations are due to the cross-sectional research design used; therefore, we cannot express any causal inferences from the obtained results.

9. Conclusions

The data obtained is relevant both for those who practice handball and their coaches. Often, young aged Romanian sportsmen have the tendency of being competitive, using an external referential, either the opinion of the coach, or the motor qualities of other fellow players. They leave for a second level the internal referential, obtaining a state of well-being and the development of capacities by relating to their own self. Thus, they tend to experience a high level of anxiety and mistrust in their own forces, which visibly diminishes sport performances. They are aware to a too small extent that internal limits are greater than external limits. The competition with one's own self, accepting the level of performance as a departing point in the sport evolution, diminishing internal limitations, and the belief in one's own abilities to succeed, together with a good relation with one's self, giving conscious, constant attention, not only to the exterior environment but also to inner feelings, experienced emotions, and their effects on the body, are indicators of sport performance.

Attaining a state of flux has short term and long-term effects in preparing sportsmen, the optimum dosing of resources allowing openness towards new possibilities. Our results are consistent with some studies [1,57–59], even if the studies did not have handball players as respondents. One of the possible practical applications of this research would be to improve the training programme by introducing components of mindfulness and self-regulation strategies. We also recommend that, at the end of each training session, the coach use ways to facilitate the development of these two essential components in obtaining sports performance.

Future researches could focus on longitudinal studies that investigate the influence of self-regulation and practising mindfulness on sport performance. Another research direction could be exploring the methods of formation, by teaching these important competencies in fostering performance.

Author Contributions: Conceptualization: D.P. and L.-G.T.; Formal analysis: V.M., T.-M.I. and L.-G.T.; Investigation: V.M.; Methodology: V.M. and T.-M.I.; Resources: D.P., T.-M.I. and L.-G.T.; Software: D.P.; Supervision: T.-M.I.; Validation: L.-G.T.; Writing – original draft: D.P. and V.M.; Writing – review & editing: D.P., T.-M.I. and L.-G.T. All authors have read and agreed to the published version of the manuscript.

Funding: This research received no external funding.

Acknowledgments: Acknowledgements to NGO Handbal Veraflor management board and Irimia Marin Viorel, president of NGO Sporting Ghimbav.

Conflicts of Interest: No potential conflict of interest was reported by the authors.

References

1. Bartulovic, D.; Young, B.W.; McCardle, L.; Baker, J. Can athletes' reports of self-regulated learning distinguish deliberate practice from physical preparation activity? *J. Sports Sci.* **2018**, *36*, 2340–2348. [CrossRef] [PubMed]
2. Noetel, M.; Ciarrochi, J.; Van Zanden, B.; Lonsdale, C. Mindfulness and acceptance approaches to sporting performance enhancement: A systematic review. *Int. Rev. Sport Exerc. Psychol.* **2017**, *12*, 1–37. [CrossRef]
3. Bühlmayer, L.; Birrer, D.; Röthlin, P.; Faude, O.; Donath, L. Effects of mindfulness practice on performance-relevant parameters and performance outcomes in sports: A meta-analytical review. *Sports Med.* **2017**, *47*, 2309–2321. [CrossRef] [PubMed]

4. Wolff, W.; Bieleke, M.; Hirsch, A.; Wienbruch, C.; Gollwitzer, P.M.; Schüler, J. Increase in prefrontal cortex oxygenation during static muscular endurance performance is modulated by self-regulation strategies. *Sci. Rep.* **2018**, *8*, 1–10. [CrossRef]

5. Toering, T.T.; Elferink-Gemser, M.T.; Jordet, G.; Visscher, C. Self-regulation and performance level of elite and non-elite youth soccer players. *J. Sports Sci.* **2009**, *27*, 1509–1517. [CrossRef]

6. Kaufman, K.A.; Glass, C.R.; Pineau, T.R. *Mindful Sport Performance Enhancement: Mental Training for Athletes and Coaches*; American Psychological Association: Washington, DC, USA, 2018.

7. Gross, M.; Moore, Z.E.; Gardner, F.L.; Wolanin, A.T.; Pess, R.; Marks, D.R. An empirical examination comparing the mindfulness-acceptance-commitment approach and psychological skills training for the mental health and sport performance of female student athletes. *Int. J. Sport Exerc. Psychol.* **2018**, *16*, 431–451. [CrossRef]

8. Harwood, C.G.; Thrower, S.N. Performance Enhancement and the Young Athlete: Mapping the Landscape and Navigating Future Directions. *Kinesiol. Rev.* **2019**, *8*, 171–179. [CrossRef]

9. Cellar, S.; Young, F.; Adair, H.; Twichell, E.; Arnold, K.A.; Royer, K.; Denning, B.L.; Riester, D. Trait goal orientation, self-regulation, and performance: A meta-analysis. *J. Bus. Psychol.* **2011**, *26*, 467–483. [CrossRef]

10. Iani, L.; Lauriola, M.; Chiesa, A.; Cafaro, V. Associations between mindfulness and emotion regulation: The key role of describing and nonreactivity. *Mindfulness* **2019**, *10*, 366–375. [CrossRef]

11. Stocker, E.; Englert, C.; Seiler, R. Mindfulness and self-control in sport. *J. Sport Exerc. Psychol.* **2017**, *39*, S317–S318.

12. Josefsson, T.; Ivarsson, A.; Gustafsson, H.; Stenling, A.; Lindwall, M.; Tornberg, R.; Böröy, J. Effects of mindfulness-acceptance-commitment (MAC) on sport-specific dispositional mindfulness, emotion regulation, and self-rated athletic performance in a multiple-sport population: An RCT study. *Mindfulness* **2019**, *10*, 1518–1529. [CrossRef]

13. Crivelli, D.; Balconi, M. Enhancing self-awareness and self-regulation to improve individual and collective joint-action in sports: Preliminary findings from a two-step study. In *Joint Action Meeting VIII.*; InTo Brain: Genova, Italy, 2019.

14. Zimmerman, B.J. Goal Setting: A Key Proactive Source of Academic Self-Regulation. In *Motivation and Self-Regulated Learning: Theory, Research, and Applications*; Schunk, D.H., Zimmerman, B.J., Eds.; Erlbaum: New York, NY, USA, 2008; pp. 267–295.

15. Zimmerman, B.J. Development and adaptation of expertise: The role of self-regulatory processes and beliefs. In *The Cambridge Handbook of Expertise and Expert Performance*; Ericsson, K.A., Charness, N., Feltovich, P.J., Hoffman, R.R., Eds.; Cambridge University Press: New York, NY, USA, 2006; pp. 705–722.

16. Kolovelonis, A.; Goudas, M.; Hassandra, M.; Dermitzaki, I. Self-regulated learning in physical education: Examining the effects of emulative and self-control practice. *Psychol. Sport Exerc.* **2012**, *13*, 383–389. [CrossRef]

17. Kolovelonis, A.; Goudas, M.; Dermitzaki, I. The effect of different goals and self-recording on self-regulation of learning a motor skill in a physical education setting. *Learn. Instr.* **2011**, *21*, 355–364. [CrossRef]

18. McCardle, L.; Young, B.W.; Baker, J. Self-regulated learning and expertise development in sport: Current status, challenges, and future opportunities. *Int. Rev. Sport Exerc. Psychol.* **2019**, *12*, 112–138. [CrossRef]

19. Kabat-Zinn, J. *Whereever You Go There You are*; Hachette Books: New York, NY, USA, 1994.

20. Bishop, S.R. What do we really know about mindfulness-based stress reduction? *Psychosom. Med.* **2002**, *64*, 71–83. [CrossRef]

21. Cavicchioli, M.; Movalli, M.; Maffei, C. Difficulties with emotion regulation, mindfulness, and substance use disorder severity: The mediating role of self-regulation of attention and acceptance attitudes. *Am. J. Drug Alcohol Abuse* **2019**, *45*, 97–107. [CrossRef]

22. Ennis, C.D. Educating students for a lifetime of physical activity: Enhancing mindfulness, motivation, and meaning. *Res. Q. Exerc. Sport* **2017**, *88*, 241–250. [CrossRef]

23. Brockman, R.; Ciarrochi, J.; Parker, P.; Kashdan, T. Emotion regulation strategies in daily life: Mindfulness, cognitive reappraisal and emotion suppression. *Cogn. Behav. Ther.* **2017**, *46*, 91–113. [CrossRef]

24. Hill, R.J.; McKernan, L.C.; Wang, L.; Coronado, R.A. Changes in psychosocial well-being after mindfulness-based stress reduction: A prospective cohort study. *J. Man. Manip. Ther.* **2017**, *25*, 128–136. [CrossRef]

25. Pineau, T.R.; Glass, C.R.; Kaufman, K.A.; Minkler, T.O. From losing record to championship season: A case study of mindful sport performance enhancement. *J. Sport Psychol. Action* **2019**, *10*, 1–11. [CrossRef]

26. Griffith, R.L.; Steelman, L.A.; Wildman, J.L.; LeNoble, C.A.; Zhou, Z.E. Guided mindfulness: A self-regulatory approach to experiential learning of complex skills. *Theor. Issues Ergon. Sci.* **2017**, *18*, 147–166. [CrossRef]

27. Birrer, D.; Röthlin, P.; Morgan, G. Mindfulness to enhance athletic performance: Theoretical considerations and possible impact mechanisms. *Mindfulness* **2012**, *3*, 235–246. [CrossRef]

28. Josefsson, T.; Ivarsson, A.; Lindwall, M.; Gustafsson, H.; Stenling, A.; Böröy, J.; Mattsson, E.; Carnebratt, J.; Sevholt, S.; Falkevik, E. Mindfulness mechanisms in sports: Mediating effects of rumination and emotion regulation on sport-specific coping. *Mindfulness* **2017**, *8*, 1354–1363. [CrossRef] [PubMed]

29. Yook, Y.S.; Kang, S.J.; Park, I.K. Effects of physical activity intervention combining a new sport and mindfulness yoga on psychological characteristics in adolescents. *Int. J. Sport Exerc. Psychol.* **2017**, *15*, 109–117. [CrossRef]

30. Wagner, H.; Finkenzeller, T.; Würth, S.; Von Duvillard, S.P. Individual and team performance in team-handball: A review. *J. Sports Sci. Med.* **2014**, *13*, 808.

31. Ziv, G.; Lidor, R. Physical characteristics, physiological attributes, and on-court performances of handball players: A review. *Eur. J. Sport Sci.* **2009**, *9*, 375–386. [CrossRef]

32. Mohamed, H.; Vaeyens, R.; Matthys, S.; Multael, M.; Lefevre, J.; Lenoir, M.; Philippaerts, R. Anthropometric and performance measures for the development of a talent detection and identification model in youth handball. *J. Sports Sci.* **2009**, *27*, 257–266. [CrossRef]

33. Michalsik, L.B.; Madsen, K.; Aagaard, P. Match performance and physiological capacity of female elite team handball players. *Int. J. Sports Med.* **2014**, *35*, 595–607. [CrossRef]

34. Michalsik, L.B.; Aagaard, P.; Madsen, K. Technical activity profile and influence of body anthropometry on playing performance in female elite team handball. *J. Strength Cond. Res.* **2015**, *29*, 1126–1138. [CrossRef]

35. Moore, Z.E.; Bonagura, K. Current opinion in clinical sport psychology: From athletic performance to psychological well-being. *Curr. Opin. Psychol.* **2017**, *16*, 176–179. [CrossRef]

36. Bjork, R.A.; Dunlosky, J.; Kornell, N. Self-regulated learning: Beliefs, techniques, and illusions. *Annu. Rev. Psychol.* **2013**, *64*, 417–444. [CrossRef] [PubMed]

37. Bartulovic, D.; Young, B.W.; Baker, J. Self-regulated learning predicts skill group differences in developing athletes. *Psychol. Sport Exerc.* **2017**, *31*, 61–69. [CrossRef]

38. Cox, A.E.; Ullrich-French, S.; French, B.F. Validity evidence for the state mindfulness scale for physical activity. *Meas. Phys. Educ. Exerc. Sci.* **2016**, *20*, 38–49. [CrossRef]

39. Beaton, D.E.; Bombardier, C.; Guillemin, F.; Ferraz, M.B. Guidelines for the process of cross-cultural adaptation of self-report measures. *Spine* **2000**, *25*, 3186–3191. [CrossRef] [PubMed]

40. Toering, T.; Elferink-Gemser, M.; Jonker, L.; van Heuvelen, M.; Visscher, C. Measuring self-regulation in a learning context: Reliability and validity of the self-regulation of learning self-report scale (SRL-SRS). *Int. J. Sport Exerc. Psychol.* **2012**, *10*, 24–38. [CrossRef]

41. Pitkethly, A.J.; Lau, P.W. Reliability and validity of the short Hong Kong Chinese Self-Regulation of Learning Self-Report Scale (SRL-SRS-C). *Int. J. Sport Exerc. Psychol.* **2016**, *14*, 210–226. [CrossRef]

42. Papalia, D.E.; Olds, S.W.; Feldman, R.D. *Human Development*; McGraw-Hill: New York, NY, USA, 2007.

43. Lemel, H.; Granér, S.; Apitzsch, E. The Effect of Acceptance-Based Intervention on Performance among Female Junior Handball Players. Master's Thesis, Lund University, Lund, Sweden. Available online: https: //www.researchgate.net/profile/Henrik_Lemel/publication/258363226_The_Effect_of_Acceptance-Based_ Intervention_on_Performance_among_Female_Junior_Handball_Players/links/004635280fc1c18f5a000000/ The-Effect-of-Acceptance-Based-Intervention-on-Performance-among-Female-Junior-Handball-Players.pdf (accessed on 14 February 2020).

44. Graham, J.D.; Li, Y.C.; Bray, S.R.; Cairney, J. Effects of cognitive control exertion and motor coordination on task self-efficacy and muscular endurance performance in children. *Front. Hum. Neurosci.* **2018**, *12*, 379. [CrossRef]

45. Gagnon-Dolbec, A.; McKelvie, S.J.; Eastwood, J. Feedback, sport-confidence and performance of lacrosse skills. *Curr. Psychol.* **2019**, *38*, 1622–1633. [CrossRef]

46. Kang, Y.; Rahrig, H.; Eichel, K.; Niles, H.F.; Rocha, T.; Lepp, N.E.; Britton, W.B. Gender differences in response to a school-based mindfulness training intervention for early adolescents. *J. Sch. Psychol.* **2018**, *68*, 163–176. [CrossRef]

47. Montroy, J.J.; Bowles, R.P.; Skibbe, L.E.; McClelland, M.M.; Morrison, F.J. The development of self-regulation across early childhood. *Dev. Psychol.* **2016**, *52*, 1744. [CrossRef]

48. Nakanishi, M.; Yamasaki, S.; Endo, K.; Ando, S.; Morimoto, Y.; Fujikawa, S.; Hiraiwa-Hasegawa, M. The association between role model presence and self-regulation in early adolescence: A cross-sectional study. *PLoS ONE* **2019**, *14*. [CrossRef]
49. Masters, R.; Maxwell, J. The theory of reinvestment. *Int. Rev. Sport Exerc. Psychol.* **2008**, *1*, 160–183. [CrossRef]
50. Martin, J.J.; Gill, D.L. The relationships among competitive orientation, sport-confidence, self-efficacy, anxiety, and performance. *J. Sport Exerc. Psychol.* **1991**, *13*, 149–159. Available online: http://digitalcommons.wayne.edu/coe_khs/1 (accessed on 21 August 2019). [CrossRef]
51. Iwatsuki, T.; Van Raalte, J.L.; Brewer, B.W.; Petitpas, A.; Takahashi, M. Relations among reinvestment, self-regulation, and perception of choking under pressure. *J. Hum. Kinet.* **2018**, *65*, 281–290. [CrossRef] [PubMed]
52. Kitsantas, A.; Zimmerman, B.J. Comparing self-regulatory processes among novice, non-expert, and expert volleyball players: A microanalytic study. *J. Appl. Sport Psychol.* **2002**, *14*, 91–105. [CrossRef]
53. Ullrich-French, S.; González Hernández, J.; Hildago Montesinos, M.D. Validity Evidence for the Adaptation of the State Mindfulness Scale for Physical Activity (SMS-PA) in Spanish Youth. *Psicothema* **2017**, *29*, 119–125. [CrossRef]
54. Niemiec, R.M. *Mindfulness and Character Strengths*; A practical guide to flourishing; Hogrefe Publishing: Boston, MA, USA, 2013.
55. Kohls, N.; Sauer, S.; Walach, H. Facets of mindfulness–Results of an online study investigating the Freiburg mindfulness inventory. *Personal. Individ. Differ.* **2009**, *46*, 224–230. [CrossRef]
56. Boekaerts, M.; Corno, L. Self-regulation in the classroom: A perspective on assessment and intervention. *Appl. Psychol.* **2005**, *54*, 199–231. [CrossRef]
57. Bernier, M.; Thienot, E.; Codron, R.; Fournier, J.F. Mindfulness and acceptance approaches in sport performance. *J. Clin. Sport Psychol.* **2009**, *3*, 320–333. [CrossRef]
58. Chiviacowsky, S. Self-controlled practice: Autonomy protects perceptions of competence and enhances motor learning. *Psychol. Sport Exerc.* **2014**, *15*, 505–510. [CrossRef]
59. Baltzell, A.L. Mindfulness and performance. In *Mindfulness in Positive Psychology*; Ivtzan, I., Lomas, T., Eds.; House Routledge: New York, NY, USA, 2016; pp. 74–89.

 *sustainability*

Review

Associations of Physical Activity and Sedentary Behaviour Assessed by Accelerometer with Body Composition among Children and Adolescents: A Scoping Review

Emanuela Gualdi-Russo [1], Natascia Rinaldo [1,*], Stefania Toselli [2,*] and Luciana Zaccagni [1,3]

[1]   Department of Neuroscience and Rehabilitation, Faculty of Medicine, Pharmacy and Prevention,
     University of Ferrara, 44121 Ferrara, Italy; emanuela.gualdi@unife.it (E.G.-R.); luciana.zaccagni@unife.it (L.Z.)
[2]   Department of Biomedical and Neuromotor Science, University of Bologna, 40126 Bologna, Italy
[3]   Center of Sport and Exercise Sciences, University of Ferrara, 44123 Ferrara, Italy
*   Correspondence: natascia.rinaldo@unife.it (N.R.); stefania.toselli@unibo.it (S.T.)

Received: 2 December 2020; Accepted: 28 December 2020; Published: 31 December 2020

Abstract: The possible adverse health effects of reduced physical activity (PA) on children and adolescents have been extensively documented as a result of the global obesity epidemic. However, the research has sometimes led to controversial results, due to the different methods used for the assessment of PA. The main aim of this review was to evaluate the association between PA and body composition parameters based on quantitative PA studies using the same equipment (Actigraph accelerometer) and cutoffs (Evenson's). A literature review was undertaken using PUBMED and Scopus databases. Subjects aged 6–15 were considered separately by sex. Weighted multiple regression analyses were conducted. From the analysis of fourteen selected articles, it emerged that 35.7% did not evaluate the association of sedentary time (ST) and moderate-to-vigorous physical activity (MVPA) with body composition, while the remaining 64.3% found a negative association of MVPA with BMI and fat mass with different trends according to sex. Furthermore, only 7.1% of these studies identified a positive association between ST and fat percentage. Based on the regression analyses conducted on the literature data, ST and MVPA were found to be significant predictors of body composition parameters, in addition to age and sex. Further studies using standardized methodologies to assess PA and body composition are needed. The inclusion of sex-disaggregated data may also be crucial to understand this phenomenon and to provide stronger evidence of the determinants of body composition in order to prevent the risk of obesity.

Keywords: accelerometer; sedentary behaviour; moderate-to-vigorous physical activity; age; sex; body composition

---

1. Introduction

The contribution of physical activity (PA) to a healthy lifestyle for children and adults is well known [1,2]. The prevention of overweight and obesity is one of the main benefits of PA in both children and adolescents. The role played by the PA should be well understood, considering that obesity prevalence has risen worldwide, even in rural areas [3,4]. In fact, previous studies have shown inconsistent results regarding the association between PA and adiposity or body mass index (BMI): increased activity is associated with a lower BMI [5,6], although PA interventions so far seem ineffective in improving the body mass index or body composition of children [7,8]. One possible explanation is that such interventions do not increase children's activity enough to affect adiposity.

According to Wilks et al. [9], PA would not be a major predictive factor for the changing of adiposity among children and adolescents. This is in contrast to the results of Kwon et al. [10], which are based on accelerometry data (moderate-to-vigorous physical activity (MVPA)) and accurate adiposity indicators (fat percentage (%F) and fat mass index) in a longitudinal sample. This latest study concluded that maintaining 45 min of accelerometry-measured daily MVPA throughout childhood and adolescence constitutes an active lifestyle to prevent obesity in young adulthood. Confirming these outcomes, greater values of MVPA (assessed by accelerometry) were associated with a lower risk of obesity in children aged 9–11 years in the study by Katzmarzyk et al. [11]. Differences in PA measurements and adiposity indicators likely contributed to such inconsistent results [9,12], hampering the comparisons of studies.

The definition of PA intensity is complex and confusing due to the different approaches used to evaluate it [13]. Many previous studies are based on the assessment of PA through questionnaires; these represent a weak substitute for objectively measured PA [14]. An overview of different PA assessment techniques can be found in McClung et al. [13]. In this review, we intend to focus on quantitative assessments and in particular on accelerometry.

The accelerometer measures the acceleration of the body, providing an objective assessment of PA intensity and total body movement (counts·min^{-1}) [15]. This device enables a large amount of data to be collected, but it has some limitations.

Considering the data from the literature, particular caution is needed as significant differences have been found between results obtained using different accelerometer brands [16]. Among the various accelerometers currently available, Actigraph (Pensacola, FL, USA) is one of the most widely used; the majority of experimental studies on PA and sedentary time (ST) use this type of accelerometer, with different types of signals (uniaxial or triaxial) [17]. Differences between uniaxial and triaxial accelerometry assessments are deemed negligible when applied to monitor daily routine, but not in specific sports [18–20].

In addition to the possible different outcomes from different brands of devices, there are also those related to the different cutoff points used to define the levels of PA. Trost et al. [15] assessed the classification accuracy of five different sets of cutpoints [21–25] for the Actigraph accelerometer for children and adolescents, concluding that the Evenson et al. [25] cutoff points are "the best overall performer across all intensity levels". In particular, an ST cutpoint of 100 cpm showed excellent classification accuracy, as did a cutpoint of ≥2296 cpm for moderate-to-vigorous physical activity (MVPA), proposed by Evenson et al. [25].

A further element of confusion is that many studies, even recent ones, evaluate the association of PA with body composition in mixed-sex samples. It is known that boys spend more time in MVPA than girls [26], and girls show a greater amount of fat than boys even during pre-puberty, with increasing differences in fat mass (FM) and fat-free-mass (FFM) during puberty [27,28].

In consideration of the inconsistent and contradictory results for the association between PA and body composition parameters in children and adolescents, we collected only the studies using: (i) a homogeneous methodology to evaluate PA objectively, using in particular the Actigraph accelerometer and PA levels according to Evenson et al. cutoffs; (ii) data of PA and body composition parameters reported separately by sex. Through this strategy, we provided evidence on this topic based both on the existing literature, summarising the most relevant information, and on the results of statistical analyses carried out on the data collected by the selected studies. Therefore, this study supplies an overview of studies from literature investigating PA levels and body composition parameters during childhood and adolescence, on the other. In addition, it assesses the importance of sex, age, ST and MVPA as determinants of BMI, %F, and fat free mass index (FFMI).

This study hence provides a contribution to the current knowledge, giving useful insights into how PA and body composition relate to one another and how to develop future research strategies.

2. Materials and Methods

2.1. Searching Strategy

This is a scoping review [29] of existing studies involving PA and body composition in school-aged children and adolescents. This review was carried out in accordance with PRISMA guidelines [30,31] (Figure 1). The articles published between January 2010 and January 2020 were systematically searched on PUBMED/MEDLINE and Scopus.

The search strategy used all the combinations with the following words: ("accelerometer" OR "Actigraph") AND ("body composition" OR "adiposity" OR "fat mass" OR "fatness" OR "fat free mass").

Two reviewers (E.G.-R. and N.R.) selected the relevant studies separately based on the titles and abstracts. Then, all the authors independently reviewed the full text of the selected studies to decide on their final suitability according to the inclusion and exclusion criteria. The references of selected studies were reviewed to search for further studies to be included in this research.

The following information was collected from each included article: authors, year of publication, study name (if present) and design, sample size, mean age and country of subjects included in the study, ST, MVPA and anthropometric parameters (stature, weight, BMI, %F, FM, FFM and FFMI).

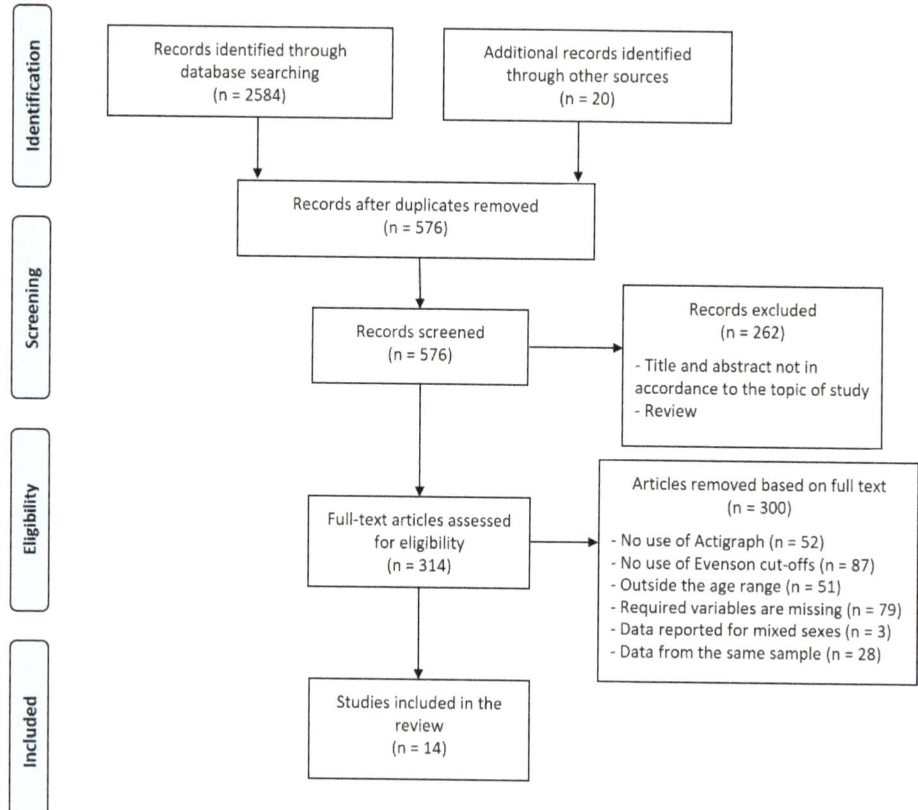

Figure 1. Flowchart detailing the search for eligible studies.

2.2. Criteria for Inclusion and Exclusion

Articles were selected through database searches using the following inclusion criteria: (1) PA intensity detected with an Actigraph accelerometer; (2) classification of PA intensity by cutoffs in accordance with Evenson et al. [25]; (3) mean age of the sample in the 6–15 year range; (4) the minimum variables required to enter into the review were ST, MVPA, %F and FFM; (5) data were reported separately by sex. The BMI and FFMI values within the baseline characteristics of participants were deemed important, but not decisive. If some of these characters were not reported, the study was nevertheless included when missing data could be obtained by calculation from the other data reported by the authors (e.g., MVPA = moderate PA (MPA) + vigorous PA (VPA); %F = (FM/weight) * 100; etc.).

In addition, we included longitudinal studies collecting only baseline data. When they did not include the chosen variables or the age of the participants was out of the chosen range, we selected the first useful data from subsequent surveys.

In the case of several articles referring to subsamples taken from the same main study, we chose the most informative one on the basis of included variables and, as the next criterion, the study with the largest subsample size.

Exclusion criteria were the following: (1) studies not published in English; (2) studies using only pedometer/step counters; (3) studies carried out on subjects suffering from medical pathologies; (4) studies conducted on athletes or subjects examined for accelerometry during motor exercises or performance tests; (5) studies reporting results from mixed-sex samples; (6) studies concerning preschool children, adults or the elderly. Reviews, editorials, and study protocols were excluded. Some articles were not included for more than one reason. Figure 1 shows the main reasons; a mixed-sex data presentation was the most frequent additional reason.

2.3. Examined Variables

First, we chose ST, namely the time spent in sedentary behaviour, and MVPA, a category of activity intensity. ST and MVPA (or, when missing, moderate physical activity (MPA) and vigorous physical activity (VPA)), were objectively determined by an Actigraph accelerometer for several consecutive days using specific cutoff points (ST $\leq$ 100 cpm, MVPA $\geq$ 2296 cpm or MPA = 2296–4011 cpm and VPA $\geq$ 4012 cpm) [15,25]. ST and PA were reported as ST and MVPA (or MPA and VPA) per day (min/day).

Among the anthropometric characters and indices, we collected data on stature, weight, BMI, %F, FM, FFM and FFMI. The BMI was determined as follows: weight (kg)/ stature squared (m^2). FFMI was determined as follows: FFM (kg)/stature squared (m^2).

Four methods were used to assess the body composition parameters in the eligible studies: anthropometry (%F was estimated by skinfolds through specific equations); bio-impedance analysis (BIA); dual energy X-ray absorptiometry (DXA) (total body fat mass was derived from the DXA scan images); deuterium dilution method (DLW). The evaluation method used in the analysed studies is given in Table 1.

2.4. Statistical Analysis

From the 14 studies examined, we collected the outcomes relative to 6558 apparently healthy subjects (3179 males, 3379 females), which were used for statistical purposes.

To test the association between a single continuous dependent variable (BMI or %F or FFMI) and several independent variables (sex, age, ST and MVPA), multiple linear regression models were used. The "weight" option to adjust the contribution of each study in proportion to their sample size to the outcome of the multiple regression was applied in this analysis. Multicollinearity between variables was assessed through variance inflation factors (VIFs), with values less than 10 considered acceptable [45].

All statistical analyses were carried out using STATISTICA for Windows (version 11.0, StatSoft, Tulsa, OK, USA). The level of significance was set at $p \leq 0.05$.

Table 1. Characteristics of the included studies: name and design, information about Actigraph accelerometer, and association between physical activity (PA) and/or sedentary time (ST) and body composition parameters.

Reference	Study Name and Design	Accelerometer: Model, Location, Duration	Conclusions on the Association of PA or ST with Body Composition
Ben Jemaa et al., 2018 [32]	Cross-sectional study	Actigraph GT3X+ on the right hip at least for 3 weekdays and 1 weekend day	MVPA was negatively correlated with %Fat (mixed sexes)
Benitéz-Porres et al., 2016 [33]	Cross-sectional study	Actigraph GT3X on the right hip. Only participants with ≥4 complete days, including 1 weekend day, were included. A day was considered valid if it contained ≥10 h of wear time for weekdays and ≥8 h for weekend days	not assessed
Bernhardsen et al., 2019 [34]	Part of Norwegian Mother and Child Cohort Study (MoBa), a prospective population-based cohort study. Data from a sub-cohort of 1603 participants born between 2002 and 2004	Actigraph GT3X+ on right hip of all participants providing at least one valid day	MVPA was negatively associated with BMI and FM only in boys
Chaput et al., 2012 [35]	QUebec Adiposity and Lifestyle InvesTigation in Youth (QUALITY) Cross-sectional study	Actigraph LS 7164 on the right hip. A valid day was defined as 10 or more hours of monitor wear time; respondents with four or more valid days were retained for analyses	MVPA was negatively associated with %Fat (mixed sexes). No association for ST.
Diouf et al., 2016 [26]	Cross-sectional study	Actigraph GT3X+ triaxial worn on the right mid-axilla line at the level of the iliac crest for at least four valid days of data, including one weekend day with greater than 10 h/day of wear time. A valid day was defined as recording at least 600 min of measured wear time between 07:00 a.m. to bedtime 22:59.	not assessed
Ferrari et al., 2015 [36]	Part of the International Study of Childhood Obesity, Lifestyle and the Environment (ISCOLE), a multinational cross-sectional study conducted in twelve countries (Australia, Brazil, Canada, China, Colombia, Finland, India, Kenya, Portugal, South Africa, United Kingdom, and United States). Here there are data only from Brazilian sample	Actigraph GT3X+ worn at the hip on an elasticized belt, on the right midaxillary line, at least for 4 days (including at least one weekend day) with at least 10 h/day of waking wear time	MVPA was negatively associated with BMI and %Fat in males (VPA in females). No association for ST.
Gába et al., 2017 [37]	Cross-sectional study	Actigraph was worn at the hip for a minimum of 10 h of wearing time per day and at least 4 days including one weekend day	MVPA was negatively associated with Fat Mass, %Fat, Fat Mass Index only in girls (VPA with the same traits in boys)

Table 1. *Cont.*

Reference	Study Name and Design	Accelerometer: Model, Location, Duration	Conclusions on the Association of PA or ST with Body Composition
Hallal et al., 2013 [38]	A subsample of the 1993 Pelotas (Brazil) Birth Cohort study at the mean age of 13.3 years. Cross-sectional	Actigraph GT1M worn on the left side of the hip for 6 days. Days with <600 min of registered data and periods of time above 60 min of consecutive zero counts were excluded	not assessed
Herrmann et al., 2015 [39]	IDEFICS study (Identification and prevention of dietary- and lifestyle-induced health effects in children and infants), a prospective population-based cohort study of children from 7 European countries	ActiTrainer or GT1M Actigraph uniaxial accelerometers worn on right hip for three consecutive days, including 1 weekend day, for at least 6 h per day. Both types of accelerometers have been observed to measure comparable MVPA levels (except for lower PA levels).	not assessed
Janz et al., 2017 [40]	Part of Iowa Bone Development Study- a longitudinal study investigating bone health and body composition in seven measurement waves at approximate ages 5 years (yr) (wave 1), 8 yr (wave 2) -used in this review-, 11 yr (wave 3), 13 yr (wave 4), 15 yr (wave 5), 17 yr (wave 6), and 19 yr (wave 7)	Actigraph uniaxial accelerometers model 7164 in waves 1 to 4 (used in this study), model GT1m in wave 5, and model GT3x+ in waves 6 and 7 for at least 10 h/day and a minimum of 3 d within 15 months of the DXA scan, worn at the hip on the midaxillary line	MVPA was negatively associated with Fat Mass in both males and females. No association for ST.
Joensuu et al., 2018 [41]	Part of research related to Finnish Schools ' on the Move program (LIKES Research Centre for Physical Activity and Health)-longitudinal- 2013	Actigraph GT3X+ at the hip, for at least 2 weekdays and 1 weekend day, 17 h/day.	not assessed
McCormack et al., 2016 [42]	Cross-sectional study	Actigraph G3TX+ accelerometers worn on the right hip for at least 3 valid weekdays and 1 valid weekend day. One day was considered valid if the child had a minimum of 10 h of wear time during waking hours.	MVPA was negatively correlated with %Fat in mixed sexes and boys (no in females). ST was positively correlated with %Fat in mixed sexes and boys (no in females).
Santos-Magalhaes et al., 2015 [43]	Cross-sectional study	Actigraph GT3X worn on the right hip for at least 2 weekdays and 1 weekend day. A minimum recording of 8 h/day was the criteria to accept daily PA	MVPA was significantly greater in normal weight than in overweight and obese. ST did not differ.
Sardinha et al., 2017 [44]	A school-based cluster randomized controlled trial (clinical trial registry: ISRCTN76013675) to evaluate the impact of an intervention in childhood obesity between 2010 and 2011- cross-sectional and prospective study	GT1M Actigraph worn on the right hip for at least three days of recording (two weekdays and one weekend day). A minimum of 600 min was required for inclusion in the analysis	MVPA was negatively associated with Fat Mass (mixed sexes). No association for ST.

3. Results

3.1. Description of Included Studies

A total of 2604 records were identified: 2584 through database searches and 20 additional records by reference list searches. After removing duplicates, 576 records were screened; 262 were excluded by reading titles and abstracts, so the full text of 314 articles was assessed for eligibility. 300 articles were removed according to inclusion and exclusion criteria, leading to the inclusion of 14 original studies. Figure 1 shows a flowchart of the process of selecting articles.

Table 1 shows study design, type of accelerometer used to assess PA, and association of MVPA or ST with body composition parameters. Five out of fourteen studies did not analyse the association between MVPA and ST with body composition parameters this [26,33,38,39,41]. Three of the remaining nine studies found a negative association of MVPA with BMI and FM only in boys [34,36,42], one only in girls [37], one in both boys and girls [40] and three in mixed sexes [32,35,44]. Finally, the residual study observed higher MVPA values in normal weights than in overweight/obese children [43]. No association was shown between ST and body composition parameters, except in the study of McCormack et al. [42], where ST was positively correlated with %F in boys and mixed sexes.

Tables 2 and 3 show the characteristics of the studies included in the review for males and females, respectively, according to sample size, age, country, ST, MVPA, stature, weight, BMI, and body composition parameters (%F, FM, FFM and FFMI).

For the research focused on males, three studies included subjects with a mean age between 8 and 9 years [39,40,43], five with a mean age between 9 and 10 [26,32,35,37,44], four between 10 and 11 [33,34,36,42], two between 12 and 13 [38,41] and one study with a mean age of 15.4 [33]. For the research focused on females, there were three studies with a mean age range between 8 and 9 [39,40,43], six between 9 and 10 [26,32,35,37,42,44], three between 10 and 11 [33,34,36] and one study each for mean ages of 12.4 [41], 13.1 [38] and 15.2 years [33].

Most of the included studies were performed in developed countries. Seven studies were carried out in Western Europe [33,34,37,39,41,43,44], three in North America [35,40,42] and the remaining four studies in developing countries (two in Brazil [36,38] and two in Africa [26,32]). For research involving both sexes, 9% lived in developing countries and the remaining 91% in developed countries. Two articles were published by 2014 [35,38] and the remaining 12 were published in 2015–2019 [26,32–34,36,37,39–44].

The majority were descriptive cross-sectional studies (n = 9, 64.3%) [26,32,33,35–38,42,43]. Four studies (28.6%) were prospective cohort studies [34,39–41] and one study (7.1%) was cross-sectional and prospective [44].

With regard to the body composition assessment, six studies (42.9%) used DXA [34,35,40,42–44], four (28.6%) used BIA [36,37,39,41], three (21.4%) used DLW [26,32,38] and one (7.1%) used skinfold thicknesses [33]. Almost all the studies reported %F (mainly) and/or FM and FFM. Only one reported FFMI [41]. Of fourteen studies, half included BMI values [33,34,36,37,41,43,44].

Table 2. Characteristics of boys according to the examined studies (calculated values are shown in italics underlined). * Sweden, Germany, Hungary, Italy, Cyprus, Spain, Belgium and Estonia; § values converted from hour/day to min/day; a = BIA; b = DXA; c = DLW; d = skinfolds.

Reference	N	Age (Years)	Country	ST (min/d)	MVPA (min/d)	Stature (cm)	Weight (kg)	BMI (kg/m²)	%F	FM (kg)	FFM (kg)	FFMI (kg/m²)
Herrmann et al., 2015 [39]	1409	8.1 ± 1.2	Europe *	336 ± 90 §	48 ± 25	135.2 ± 9.0	32.5 ± 8.7	17.8	28.9	9.4	23.1 ± 4.1 a	12.6
Santos-Magalhaes et al., 2015 [43]	26	8.2 ± 1.2	Portugal	325.6 ± 79.7	60.7 ± 27.2	133.5 ± 9.7	35.8 ± 12.8	19.5 ± 4.7	31.4 ± 9.2 b	11.2	24.6	13.8
Janz et al., 2017 [40]	201	8.7 ± 0.7	USA	324 ± 66 §	58.2 ± 25.8	135 ± 7.3	33.4 ± 9.0	18.3	24.6	8.2 ± 5.8b	25.2	13.8
Ben Jemaa et al., 2018 [32]	21	9.3 ± 0.9	Tunisia	583.9	68.8 ± 17.3	137.1 ± 6.9	31.6 ± 6.2	16.8	22.7 ± 5.1 c	7.2	24.4	13.0
Chaput et al., 2012 [35]	299	9.6 ± 0.8	Canada	341 ± 82	55.9 ± 26.4	-	38.2 ± 11.6	-	24.9 ± 11.0 b	9.5	28.7	-
Gába et al., 2017 [37]	156	9.9 ± 1.2	Czech Republic	379.3 ± 48.5	46.7 ± 21.3	143.8 ± 9.4	37.4 ± 9.2	17.9 ± 2.8	16.1 ± 8.3 a	6.7 ± 5.2 a	30.7	14.9
Diouf et al., 2016 [26]	20	10 ± 1	Senegal	428 ± 89	85 ± 40	142.1 ± 6.5	34.1 ± 9.4	16.9	16.5 c	5.7 ± 1.2 c	24.7 ± 3.8 c	12.2
Sardinha et al., 2017 [44]	197	10 ± 0.6	Portugal	518.2 ± 59.2	65.9 ± 24.5	142 ± 7	37.8 ± 8.6	18.6 ± 3.5	26.7	10.1 ± 5.6b	27.7	13.7
Ferrari et al., 2015 [36]	238	10.1 ± 0.5	Brazil	492.5	71.2 ± 28	144.4 ± 7.1	41.5 ± 12.6	19.9 ± 4.7	21.3 ± 9.5 a	8.8	32.7	15.7
McCormack et al., 2016 [42]	44	10.1 ± 0.9	USA	550 ± 62	71 ± 25	141.5 ± 7.5	38 ± 8.2	19.0	29.1 ± 6.3 b	10.4 ± 4.5b	23.2 ± 3.9b	11.6
Benitéz-Porres et al., 2016 [33]	83	10.9 ± 1.2	Spain	599.3 ± 60.7	63.9 ± 14.7	145.6 ± 10.1	42.7 ± 10.8	19.9 ± 3.4	22.9 ± 11 d	9.8	32.9	15.5
Bernhardsen et al., 2019 [34]	98	10.9 ± 0.6	Norway	501 ± 61.6	75 ± 26.9	148 ± 6	39.2 ± 6.9	17.8 ± 2.5	24.7 ± 6.2 b	10 ± 4.2 b	29.4 ± 3.6 b	13.4
Joensuu et al., 2018 [41]	263	12.4 ± 1.3	Finland	485.5 ± 74.9	58.2 ± 23.3	155.2 ± 11.1	44.6 ± 11.9	18.2 ± 3.1	14.7 ± 7.8 a	6.6	38.0	15.4 ± 1.7
Hallal et al., 2013 [38]	9	12.9 ± 0.3	Brazil	660 ± 82	79	159.5 ± 10.5	51.9 ± 12.2	20.4	22.7	11.8 ± 6.8c	40.1 ± 7.6c	15.8
Benitéz-Porres et al., 2016 [33]	115	15.4 ± 1.3	Spain	643.1 ± 67.2	65.9 ± 23.3	167.6 ± 8.4	61.5 ± 13.7	21.8 ± 4.2	16.6 ± 7.6 d	10.2	51.3	18.3

Table 3. Characteristics of girls according to the examined studies (calculated values are shown in italics underlined). * Sweden, Germany, Hungary, Italy, Cyprus, Spain, Belgium and Estonia; § values converted from hour/day to min/day; a = BIA; b = DXA; c = DLW; d = skinfolds.

Reference	N	Age (Years)	Country	ST (min/d)	MVPA (min/d)	Stature (cm)	Weight (kg)	BMI (kg/m²)	%F	FM (kg)	FFM (kg)	FFMI (kg/m²)
Herrmann et al., 2015 [39]	1544	8.1 ± 1.2	Europe *	342 ± 90 §	36 ± 18	134.7 ± 9.1	32.1 ± 8.3	17.7	33.6	10.8	21.3 ± 4.0 a	11.7
Santos-Magalhaes et al., 2015 [43]	24	8.4 ± 1	Portugal	347.1 ± 66.5	52.1 ± 16.5	133.6 ± 6.8	36 ± 8.7	20.0 ± 3.5	37.3 ± 6.4 b	13.4	22.6	12.7
Janz et al., 2017 [40]	214	8.7 ± 0.6	USA	330 ± 66 §	40.6 ± 18.9	133.0 ± 6.8	32 ± 8.7	18.1	29.1b	9.3 ± 5.9	22.7	12.8
Diouf et al., 2016 [26]	22	9 ± 1	Senegal	416 ± 79	65 ± 28	137.3 ± 8.2	31.7 ± 8.6	16.8	24.2	7.5 ± 1.0 c	21.1 ± 3.5 c	11.2
Ben Jemaa et al., 2018 [32]	19	9.4 ± 1	Tunisia	498.8	49.8 ± 22.7	139.6 ± 8.3	36.1 ± 8.6	18.5	31.1 ± 8.2 c	11.2	24.9	12.8
Chaput et al., 2012 [35]	251	9.6 ± 0.9	Canada	343 ± 82	37.5 ± 17.9	-	38 ± 11.2	-	30.3 ± 10.0 b	11.5	26.5	-
Gába et al., 2017 [37]	209	9.8 ± 1.3	Czech Republic	394.4 ± 51.7	37.4 ± 16.7	142.5 ± 8.9	35.8 ± 8.8	17.4 ± 2.8	18.1 ± 7.3 a	6.9 ± 4.4 a	28.9	14.2
Sardinha et al., 2017 [44]	189	9.9 ± 0.6	Portugal	525.4 ± 67.1	53.7 ± 18.4	143 ± 7	39.7 ± 8.6	19.1 ± 3.2	30.5	12.1 ± 5.1 b	27.6	13.5
McCormack et al., 2016 [42]	43	10.0 ± 0.8	USA	580 ± 48	51 ± 22	140.0 ± 6.5	35.0 ± 7.0	17.9	31.5 ± 6.1 b	10.3 ± 4.0 b	20.7 ± 3.5 b	10.6
Ferrari et al., 2015 [36]	247	10.1 ± 0.5	Brazil	507.3 ± 66.7	48.2 ± 18.0	145.3 ± 7.7	41 ± 10.9	19.5 ± 4.1	24.7 ± 8.3 a	10.1	30.9	14.6
Benitéz-Porres et al., 2016 [33]	63	10.7 ± 1.3	Spain	609 ± 61.2	61.3 ± 12.7	143.2 ± 10.9	40 ± 12.7	19.1 ± 4.0	22.9 ± 6.6 d	9.2	30.8	15.0
Bernhardsen et al., 2019 [34]	88	11.0 ± 0.6	Norway	508.0 ± 55.6	59 ± 17.4	149 ± 8	40 ± 8.4	17.9 ± 2.3	28.1 ± 5.8 b	11.5 ± 4.6 b	28.6 ± 4.8 b	12.9
Joensuu et al., 2019 [41]	331	12.4 ± 1.3	Finland	511.0 ± 66.6	47.3 ± 18.0	155.1 ± 9.3	45.7 ± 10.2	18.8 ± 3.0	20.4 ± 6.9 a	9.3	36.4	14.8 ± 1.4
Hallal et al., 2013 [38]	16	13.1 ± 0.3	Brazil	661 ± 78	64	159 ± 5.6	51.7 ± 9.5	20.5	30.2	15.6 ± 5.8 c	37.3 ± 6.5 c	14.8
Benitéz-Porres et al., 2016 [33]	119	15.2 ± 1.4	Spain	642.7 ± 83.4	48.1 ± 18.9	160.2 ± 6	57.5	22.4 ± 4.6	21.2 ± 7.7 d	12.2	45.3	17.7

3.2. General Overview

We performed three weighted multiple linear regression models to test the simultaneous influence of age, PA amount (examined through minutes per day of ST and MVPA) and sex on FFMI, BMI and %F (Table 4). The variables used as predictors exerted a significant effect, as shown by the R^2 value, on the dependent variable considered ($p < 0.001$ in all three models). The three models, obtained using the same four independent variables, predicted respectively 78%, 62% and 71% of the overall variance in FFMI, BMI and %F. The VIF values were acceptable in all the analyses.

The outcomes of multiple linear regression analysis with FFMI as the dependent variable showed that, after controlling for other covariates, the decrease in ST and the increase in age, in addition to being a male, were associated with significantly higher FFMI values. According to the outcomes of multiple linear regression with BMI as the dependent variable, the increases in ST and age and being a male were associated with significantly higher BMI values, while the increase in MVPA was associated with significantly lower BMI values. The last regression analysis was carried out by considering %F as the dependent variable. This analysis shows that the increase in ST and the decreases in MVPA and age, in addition to belonging to the female sex, were associated with significantly increased %F.

Table 4. Predictors of FFMI, BMI and %F: results of multiple regression analyses.

Predictors	FFMI			BMI			%F		
	β	t	p	β	t	p	β	t	p
Age	0.935	111.445	0.000	0.312	18.322	0.000	−0.933	−71.602	0.000
MVPA	0.006	0.449	0.653	−0.074	−4.005	0.000	−0.059	−3.845	0.000
ST	−0.098	−5.489	0.000	0.546	23.699	0.000	0.259	14.849	0.000
Sex (male)	0.234	21.116	0.000	0.087	6.102	0.000	−0.325	−26.335	0.000
R^2	0.776			0.624			0.705		
R^2 adj	0.776			0.623			0.705		
p	<0.001			<0.001			<0.001		

In particular, MVPA was negatively associated with BMI (0.007 kg/m^2 of BMI decrement every minute per day of MVPA increment, unstandardized parameter B, not presented in Table 4) and %F (0.03 percent points of %F decrement every minute per day of MVPA increment). ST was negatively associated with FFMI (0.002 kg/m^2 of FFMI decrement every minute per day of ST increment) and positively associated with BMI and %F (0.006 kg/m^2 and 0.015 percentage points increment, respectively, every minute per day of ST increment). In addition, sex proved to be a significant predictor for all dependent variables: being a male increased FFMI and BMI (with an increment of 0.38 kg/m^2 and only 0.09 kg/m^2, respectively, in comparison to girls) and decreased %F (1.82 percentage points).

4. Discussion

This review aimed to evaluate the association between PA and body composition parameters in children and adolescents. After a detailed analysis of the 14 articles that met the criteria for inclusion, with a total sample of 6558 apparently healthy participants (48.5% males), a tendency towards a decrease in body adiposity with increasing MVPA and decreasing ST was evident.

PA is a complex and multi-dimensional behaviour associated with several health outcomes. Its correct quantification is therefore crucial for population assessment and possible interventions to be implemented. PA assessment can be done both using questionnaires and device-based techniques. As self-reported measures of PA have proved to have several limitations [46,47], accelerometry is the most used field method in free-living subjects to assess PA due to its accuracy and reliability.

In this review, we examined the association between objectively assessed PA (by means of an Actigraph accelerometer) and body composition parameters in children and adolescents using outcome data from 14 studies. Accelerometers offer the possibility of non-invasive and objective estimation of PA intensity levels using "activity counts" to classify PA as light, moderate, or vigorous intensity, in addition to ST. Accelerometers, used for approximately twenty years for children and adolescents [48], have now also become widely used in large-scale studies [49]. These devices are commonly worn on the wrist or the hip, with the latter placement ensuring better accuracy [50].

The different location of the device, the different types of accelerometers used, and the different cutoff points that define the PA level are some of the factors that make it difficult to compare results across studies.

For this reason, we decided to include in this review only the studies that assessed PA through a hip-worn accelerometer Actigraph (the most widely used device) and used the cutoffs reported by Evenson et al. [25]. It is important to underline that differences in the accelerometer cutpoints used can lead to different PA intensities [51] and that Evenson's cutoffs are highly recommended as they are deemed to provide an optimal estimate of PA in children aged 5–15 years [52].

It should be noted that several different Actigraph models have been used in the selected articles. The majority of activity counts were recorded by GT3X or GT3X+ models, but five studies used the original 7164 and the GT1M models. Recent literature shows that accelerometer counts and time spent in MVPA as recorded by the Actigraph GT1M, GT3X and GT3X+ are compatible with one another [52]. Moreover, the vertical axis counts recorded by the 7164 and GT1M models are generally deemed highly comparable [53,54], although there is no unanimous consensus [55]. Though differences between uniaxial and triaxial accelerometric counts are apparent in specific sports [56], negligible differences are seen in daily routine activity [20].

In summary, it can be asserted that different Actigraph models, especially the more recent generations of accelerometers, provide data that can be accurately and reliably compared.

Another point of discussion is the reliability of the accelerometers used to assess PA. In literature, there are some studies focused on the assessment of the inter-instrument reliability of different accelerometers, by comparing the outputs from accelerometers worn on opposite hips or by using high-precision devices to examine technical reliability over multiple trials. A review by Trost et al. [57] summarized some of the results reported by the literature on this topic up to 2004, concluding that nearly all the studies reported high precision reliability [57]. One more recent study published by Aadland and Ylvisåker [58] in 2015 described the results of the assessment of the inter-instrument reliability of the Actigraph GT3X+ in adults. The results showed that the reliability of the accelerometer, tested wearing two accelerometers simultaneously on the contralateral hips for 21 days, was very high and that the Actigraph GT3X+ accelerometer was a reliable tool for measuring PA in adults [58]. Only a few studies have been published on inter-instrument reliability in children. Trost et al. [59] calculated the reliability of two Actigraph accelerometers worn on the left and right hips on thirty children aged 10–14 years. The results indicated an excellent intraclass reliability coefficient (ICC = 0.87). Another more recent study examined the inter-instrument reliability for eight children with cerebral palsy using an Actigraph GT3X+; it concluded that this is a reliable and valid device to monitor PA during walking [60].

4.1. Trend of PA and Body Composition with Age, Sex and Population

The examined studies defined ST as <100 cpm, including any waking behaviour characterized by an energy expenditure of ≤1.5 metabolic equivalents (METs) in a sitting or reclining position [61]. MVPA was defined as a combination of moderate and vigorous physical activity, including any activity over 3 METs [62]. The children spent most of their time (min/day) in sedentary behaviours rather than in moderate/vigorous activities: from 44% [40] to 67.3% [42] in ST vs. 5.4% [37] to 9.35% [26] in MVPA. In particular, in studies focused on females only, three studies had average MVPA values that met WHO recommendations; two of these were carried out in developing countries (Senegal [26] and Brazil [38]) and only one was carried out in a developed country (Spain [33]). For studies focusing on males, the situation was better: all the studies carried out in developing countries [26,32,36,38] and approximately half of the studies carried out in developed countries [33,34,42–44] had mean values of MVPA higher than recommended.

Several studies (among the selected ones) evaluated the frequency of children within the study sample that complied with these recommendations for children and youths by the WHO. The majority of Norwegian children satisfied the recommended PA level of ≥60 min/d of MVPA [34], as did Senegalese children aged 8–11 (55% according to Diouf et al. [26]), American children aged 9–12 (55% according to McCormack et al. [42]) and Tunisian children aged 8–11 (47.5% according to Ben Jeema et al. [32]). Only 14% of Czech children aged 7–12 [37], 20.1% of European children aged 6–10 [39], 24.7% of Canadian children aged 8–10 [35], and 33.7% of Finnish children aged 9–15 [41] were found to have done at least 60 min of daily MVPA.

Studies reporting data separately by sex show higher frequencies of MVPA in males than in females, with percentages decreasing from developing to developed countries, as follows: 66.7% of Tunisian boys vs. 26.3% of girls [32]; 52.2% of Senegalese boys vs. 47.8% of girls [26]; 44.9% of Finnish boys vs. 24.2% of girls [41]; 36% of Canadians boys vs. 11% of girls [35]; 30.6% of European boys vs. 10.5% of girls [39]; 22% of Czech boys vs. 8% of girls [37]. An exception to this trend is represented by the study of McCormack et al. [42], which reported the highest frequencies of MVPA of our review in US children (75% boys and 35% girls).

The lack of an acceptable gold standard for measuring the body composition of children implies that several techniques were used, from the simple BMI to DXA. Half of the examined studies evaluated BMI as a proxy for adiposity. Moreover, other body composition parameters (%F, FM, FFM, FFMI) were assessed by plicometry (7.1%), BIA (28.6%), DLW (21.4%) and DXA (42.9%). In particular, it should be noted that only one of the 14 studies analyzed used skinfolds as a measure.

Changes in body composition parameters occur during childhood, even if less pronounced than during infancy. Sex differences in body composition, present throughout childhood, appear more pronounced during adolescence due to the effect of gonadal sex steroids. Sex-specific differences become evident during this phase of growth: females show greater FM and lower FFM than males [63,64]. Consistent with this trend, our statistical analysis on the cross-sectional data collected in the selected studies showed a decrease in %F with age, especially in males, and an increase in FFMI and BMI, in males and females, respectively. The assessment of these parameters is very important due to the association of the amount of adult visceral adipose tissue with BMI changes during adolescence, as verified by longitudinal design [65].

If we analyze the body composition of the participants divided by developed and developing countries, is clear that males from African countries (Tunisia and Senegal) have the lowest mean BMI (also %F and FFMI in Senegalese boys) among the other studies reporting data for individuals of the same age. Brazilian subjects, however, have higher mean BMI, but similar %F and FFMI, as compared to individuals from other countries. Such differences were not present in the female sex.

4.2. Association between PA and Body Composition Parameters

The impact of sedentary behaviour on health has been investigated by many studies, with particular reference to the relationship between sedentary behaviour and childhood obesity or other health outcomes [66–69]. In recent years, the excess of adiposity in developing countries and the need for intervention to increase PA were largely reported in the literature [70–72]. Ferrari et al. [73] found that self-reported measures significantly underestimated sedentary time and overestimated PA time compared to device measures. Moreover, regression models showed that self-reported PA and sedentary behaviour were not significantly associated with health outcomes, unlike objectively-measured PA and ST. Therefore, self-reported methods may underestimate the strength of some relationships between activity and body composition.

The current study adds to the existing knowledge by providing the findings of the associations between PA (objectively assessed) and body composition parameters on a very large sample of children and adolescents taken from the literature.

The results obtained using multiple regression analysis show that the strength of the association of BMI, %F, and FFMI with MVPA and ST was also influenced by age and sex. It is important to underline the effect of sex on these associations, which has often been neglected in the literature. In our study, PA assessed by accelerometry resulted in a significant predictor of body composition parameters. The regression models explained a high and significant percentage of the variance in the body composition parameters. In particular, our analyses showed that MVPA was negatively associated with BMI and %F, and ST was negatively associated with FFMI but positively associated with BMI and %F. Subjects who spent more time engaged in MVPA showed lower weight status and lower levels of adiposity, and subjects who were more sedentary had higher weight status and %F and lower FFMI.

Thus, MVPA had a significant and positive effect on adiposity and consequently on health indicators, but not on FFMI. This is likely because the examined samples were not very physically active.

From our findings, it seems that youths living in developing countries are more physically active than those living in developed ones. Since overweight/obese youth are generally less active than normal/underweight ones [26], the lower overweight/obesity prevalence in developing countries could explain these differences. This assumption is in agreement with a cross-sectional study carried out in Tunisia, where the percentage of normal-weight children who spent more than 60 min a day in MVPA (60.9%) was significantly higher than in overweight children (29.4%) (according to data tabulated by Ben Jeema et al. [32]).

However, the study by Guthold et al. [74] gave inconsistent results; there was no increase in the prevalence of insufficient activity with country income, as was seen in high-income Asia Pacific countries. It should be noted that in this review there was no data from Asian countries.

4.3. Strength and Limitations

The collection of evidence by this scoping review is valuable in supplying an overview of the literature, with the intent of identifying where evidence gaps exist in order to assist future studies. This scoping review has several strengths, which include a systematic search of literature articles and consideration of a wide range of evidence. In particular, the major strengths of this review are that it includes only research that objectively measured ST and MVPA using the same accelerometer brand placed at the hip as well as the same cutpoints for PA classification.

The main limitation is common to any scoping review: it does not involve a quality assessment. Among the other limitations of this study, there is the use of BMI as an indicator of adiposity [75,76]; its accuracy is believed to increase in accordance with the degree of body fatness [77]. The lack of standard protocols for the detection of the body composition parameters used (%F, FM, FFM and FFMI) is another limitation to this review. The interchangeability of body composition parameters obtained by different methods is controversial. The minimal differences between BIA and DXA outcomes found in subjects aged 16-18 suggest that these methods may be interchangeable at the population level according to Achamrah et al. [78]. However, these findings contrast with those obtained by Lee et al. [79], indicating an overestimation of whole body muscle mass and skeletal muscle mass assessed by BIA as compared to DXA, especially in men. Moreover, differences in the accuracy of the values obtained in the comparison between DXA and BIA using different BIA machines has also been reported [80]. Finally, differences between the results obtained from the skinfold prediction equations compared to DXA have been shown in the literature [81,82].

Some other limitations should be mentioned, including the use of cross-sectional data (or collected by us cross-sectionally from longitudinal studies for needs related to the statistical processing of data), since this type of study design makes it difficult to explore causalities. Longitudinal data are required to further investigate the association between ST, MVPA and body composition over time [83].

Consistent with the chosen inclusion/exclusion criteria, we examined only 14 studies from the literature, and our results cannot be considered representative of the entire child and adolescent population of each country. Concerning the country of the samples, the majority of included research was from developed countries (Europe and North America) and just four studies were from developing countries (South America and Africa). Therefore, the underrepresentation of developing countries and the lack of published results from Asia has to be underlined. Although there are several studies in the literature on PA evaluated with accelerometry in Asian children and adolescents, these studies employed different accelerometers (e.g., [84,85]) and/or different cutoff points (e.g., [84–86]).

We selected only those studies that used the same type of device and cutoffs, but caution should be taken in evaluating the results obtained due to other possible variations among laboratories in procedures and methods (differences in registration period protocol, non-wear time definition, epoch length, etc.) [87]. In particular, one limitation evident in the current study is that a different number of valid days (defined as hours of monitor wear time) of accelerometer wear time was used for the final analysis. Only one study required at least six valid days of accelerometer wear time to be included in the final analysis [38], suggesting a possible PA undersampling [88]. A valid day was defined as a minimum of 6 h [26] to a maximum of 17 h [41] of wearing time per day, depending on the study.

Finally, concerning the size of the samples, it should be noted that it was not homogenous within the selected studies. The study with the smallest sample was reported by Hallal et al. [38] (9 males and 16 females). Meanwhile, the study with the largest sample was published by Herrmann et al. [39] (1409 males and 1544 females). To avoid biases in the statistical analysis we weighted our results for the number of the subjects of each study. In addition, we could not dispose of data regarding the first two age classes (6–7) in the chosen age range (6–15).

5. Conclusions

This scoping review first summarized the extent and nature of the research and identified gaps in the existing literature in order to subsequently identify key research priorities in PA and body composition. Consistent methods of measurement and analysis for both PA and body composition are required to allow comparison between literature studies and to achieve definitive conclusions. Despite some limitations, the data presented here support the suggestion that lower volumes of MVPA and higher volumes of ST are associated with increases in FM, which is detrimental to health and necessitates identification of appropriate interventions to reduce sedentary time and inactivity.

This scoping review confirmed that the adoption of standardized methods and analysis protocols for the assessment of PA and body composition remains an important concern for this area in order to improve comparability between studies. In order to obtain a realistic picture of the situation for children and adolescents, we also suggest reporting the data separately by sex. Only in this way will a better understanding of the associations between sedentary behaviour and daily PA at different levels of intensity be achieved in childhood and adolescence to successfully prevent early fat accumulation in children and lower the risk of adult obesity and its consequent comorbidities. Therefore, all children and adolescents, especially girls (who are more inactive and have greater adiposity in comparison to boys), should be encouraged to partake in PA, because of obesity seems to be due mainly to decreased PA and increased sedentary behaviour.

Author Contributions: Conceptualization, E.G.-R. and L.Z.; methodology, E.G.-R. and N.R.; formal analysis, L.Z. and N.R.; investigation, E.G.-R. and N.R.; data curation, N.R., L.Z. and S.T.; writing—original draft preparation, E.G.-R., L.Z. and N.R.; writing—review and editing, E.G.-R., L.Z., N.R. and S.T.; supervision, E.G.-R.; project administration, E.G.-R. All authors have read and agreed to the published version of the manuscript.

Funding: This research received no external funding.

Institutional Review Board Statement: Not applicable.

Informed Consent Statement: Not applicable.

Data Availability Statement: Data is contained within the article.

Conflicts of Interest: The authors declare no conflict of interest.

References

1. Ha, A.S.; Ng, J.Y.Y.; Lonsdale, C.; Lubans, D.R.; Ng, F.F. Promoting physical activity in children through family-based intervention: Protocol of the "Active 1 + FUN" randomized controlled trial. *BMC Public Health* **2019**, *19*, 218. [CrossRef]

2. Zaccagni, L.; Rinaldo, N.; Bramanti, B.; Gualdi-Russo, E. Relation between lifestyle behaviors and body composition patterns among healthy young Italians: A cross-sectional study. *J. Sports Med. Phys. Fit.* **2018**, *58*, 1652–1656. [CrossRef]

3. NCD Risk Factor Collaboration (NCD-RisC). Worldwide trends in body-mass index, underweight, overweight, and obesity from 1975 to 2016: A pooled analysis of 2416 population-based measurement studies in 128 9 million children, adolescents, and adults. *Lancet* **2017**, *390*, 2627–2642. [CrossRef]

4. NCD Risk Factor Collaboration (NCD-RisC). Rising rural body-mass index is the main driver of the global obesity epidemic in adults. *Nature* **2019**, *569*, 260–264. [CrossRef] [PubMed]

5. Ness, A.R.; Leary, S.D.; Mattocks, C.; Blair, S.N.; Reilly, J.J.; Wells, J.; Ingle, S.; Tilling, K.; Smith, G.D.; Riddoch, C. Objectively measured physical activity and fat mass in a large cohort of children. *PLoS Med.* **2007**, *4*, e97. [CrossRef] [PubMed]

6. Metcalf, B.; Henley, W.; Wilkin, T. Effectiveness of intervention on physical activity of children: Systematic review and meta-analysis of controlled trials with objectively measured outcomes (EarlyBird 54). *BMJ* **2012**, *345*, e5888. [CrossRef] [PubMed]

7. Harris, K.C.; Kuramoto, L.K.; Schulzer, M.; Retallack, J.E. Effect of school-based physical activity interventions on body mass index in children: A meta-analysis. *CMAJ* **2009**, *180*, 719–726. [CrossRef] [PubMed]

8. Waters, E.; de Silva-Sanigorski, A.; Hall, B.J.; Brown, T.; Campbell, K.J.; Gao, Y.; Armstrong, R.; Prosser, L.; Summerbell, C.D. Interventions for preventing obesity in children. *Cochrane Database Syst. Rev.* **2011**, CD001871. [CrossRef] [PubMed]

9. Wilks, D.C.; Sharp, S.J.; Ekelund, U.; Thompson, S.G.; Mander, A.P.; Turner, R.M.; Jebb, S.A.; Lindroos, A.K. Objectively measured physical activity and fat mass in children: A bias-adjusted meta-analysis of prospective studies. *PLoS ONE* **2011**, *6*, e17205. [CrossRef] [PubMed]

10. Kwon, S.; Janz, K.F.; Letuchy, E.M.; Burns, T.L.; Levy, S.M. Active lifestyle in childhood and adolescence prevents obesity development in young adulthood. *Obesity* **2015**, *23*, 2462–2469. [CrossRef]

11. Katzmarzyk, P.T.; Barreira, T.V.; Broyles, S.T.; Champagne, C.M.; Chaput, J.-P.; Fogelholm, M.; Hu, G.; Johnson, W.D.; Kuriyan, R.; Kurpad, A.; et al. Physical Activity, Sedentary Time, and Obesity in an International Sample of Children. *Med. Sci. Sports Exerc.* **2015**, *47*, 2062–2069. [CrossRef] [PubMed]

12. Kelley, G.A.; Kelley, K.S.; Pate, R.R. Exercise and BMI in Overweight and Obese Children and Adolescents: A Systematic Review and Trial Sequential Meta-Analysis. *BioMed Res. Int.* **2015**, *2015*, 704539. [CrossRef] [PubMed]

13. McClung, H.L.; Ptomey, L.T.; Shook, R.P.; Aggarwal, A.; Gorczyca, A.M.; Sazonov, E.S.; Becofsky, K.; Weiss, R.; Das, S.K. Dietary Intake and Physical Activity Assessment: Current Tools, Techniques, and Technologies for Use in Adult Populations. *Am. J. Prev. Med.* **2018**, *55*, e93–e104. [CrossRef] [PubMed]
14. Treuth, M.S.; Hou, N.; Young, D.R.; Maynard, L.M. Validity and reliability of the Fels physical activity questionnaire for children. *Med. Sci. Sports Exerc.* **2005**, *37*, 488–495. [CrossRef]
15. Trost, S.G.; Loprinzi, P.D.; Moore, R.; Pfeiffer, K.A. Comparison of accelerometer cut points for predicting activity intensity in youth. *Med. Sci. Sports Exerc.* **2011**, *43*, 1360–1368. [CrossRef]
16. Duncan, S.; Stewart, T.; Bo Schneller, M.; Godbole, S.; Cain, K.; Kerr, J. Convergent validity of ActiGraph and Actical accelerometers for estimating physical activity in adults. *PLoS ONE* **2018**, *13*, e0198587. [CrossRef]
17. Wijndaele, K.; Westgate, K.; Stephens, S.K.; Blair, S.N.; Bull, F.C.; Chastin, S.F.M.; Dunstan, D.W.; Ekelund, U.; Esliger, D.W.; Freedson, P.S.; et al. Utilization and Harmonization of Adult Accelerometry Data: Review and Expert Consensus. *Med. Sci. Sports Exerc.* **2015**, *47*, 2129–2139. [CrossRef]
18. Colley, R.C.; Garriguet, D.; Janssen, I.; Craig, C.L.; Clarke, J.; Tremblay, M.S. Physical activity of Canadian children and youth: Accelerometer results from the 2007 to 2009 Canadian Health Measures Survey. *Health Rep.* **2011**, *22*, 15–23.
19. Vanhelst, J.; Béghin, L.; Duhamel, A.; Bergman, P.; Sjöström, M.; Gottrand, F. Comparison of uniaxial and triaxial accelerometry in the assessment of physical activity among adolescents under free-living conditions: The HELENA study. *BMC Med. Res. Methodol.* **2012**, *12*, 26. [CrossRef]
20. Smith, M.P.; Horsch, A.; Standl, M.; Heinrich, J.; Schulz, H. Uni- and triaxial accelerometric signals agree during daily routine, but show differences between sports. *Sci. Rep.* **2018**, *8*, 15055. [CrossRef]
21. Freedson, P.; Pober, D.; Janz, K.F. Calibration of accelerometer output for children. *Med. Sci. Sports Exerc.* **2005**, *37*, S523–S530. [CrossRef] [PubMed]
22. Puyau, M.R.; Adolph, A.L.; Vohra, F.A.; Butte, N.F. Validation and calibration of physical activity monitors in children. *Obes. Res.* **2002**, *10*, 150–157. [CrossRef] [PubMed]
23. Treuth, M.S.; Schmitz, K.; Catellier, D.J.; McMurray, R.G.; Murray, D.M.; Almeida, M.J.; Going, S.; Norman, J.E.; Pate, R. Defining accelerometer thresholds for activity intensities in adolescent girls. *Med. Sci. Sports Exerc.* **2004**, *36*, 1259–1266. [PubMed]
24. Mattocks, C.; Leary, S.; Ness, A.; Deere, K.; Saunders, J.; Tilling, K.; Kirkby, J.; Blair, S.N.; Riddoch, C. Calibration of an accelerometer during free-living activities in children. *Int. J. Pediatr. Obes.* **2007**, *2*, 218–226. [CrossRef]
25. Evenson, K.R.; Catellier, D.J.; Gill, K.; Ondrak, K.S.; McMurray, R.G. Calibration of two objective measures of physical activity for children. *J. Sports Sci.* **2008**, *26*, 1557–1565. [CrossRef]
26. Diouf, A.; Thiam, M.; Idohou-Dossou, N.; Diongue, O.; Mégné, N.; Diallo, K.; Sembène, P.M.; Wade, S. Physical activity level and sedentary behaviors among public school children in dakar (senegal) measured by paq-c and accelerometer: Preliminary results. *Int. J. Environ. Res. Public Health* **2016**, *13*, 998. [CrossRef]
27. Malina, R.M.; Koziel, S.; Bielicki, T. Variation in subcutaneous adipose tissue distribution associated with age, sex, and maturation. *Am. J. Hum. Biol.* **1999**, *11*, 189–200. [CrossRef]
28. Kirchengast, S. Gender differences in body composition from childhood to old age: An evolutionary point of view. *J. Life Sci.* **2010**, *2*, 1–10. [CrossRef]
29. Grant, M.J.; Booth, A. A typology of reviews: An analysis of 14 review types and associated methodologies. *Health Inf. Libr. J.* **2009**, *26*, 91–108. [CrossRef]
30. Moher, D.; Liberati, A.; Tetzlaff, J.; Altman, D.G. Preferred reporting items for systematic reviews and meta-analyses: The PRISMA Statement. *Open Med.* **2009**, *3*, e123–e130.
31. Tricco, A.C.; Lillie, E.; Zarin, W.; O'Brien, K.K.; Colquhoun, H.; Levac, D.; Moher, D.; Peters, M.D.J.; Horsley, T.; Weeks, L.; et al. PRISMA Extension for Scoping Reviews (PRISMA-ScR): Checklist and Explanation. *Ann. Intern. Med.* **2018**, *169*, 467–473. [CrossRef] [PubMed]
32. Ben Jemaa, H.; Mankaï, A.; Mahjoub, F.; Kortobi, B.; Khlifi, S.; Draoui, J.; Minaoui, R.; Karmous, I.; Ben Hmad, H.; Ben Slama, F.; et al. Physical Activity Level Assessed by Accelerometer and PAQ-C in Tunisian Children. *Ann. Nutr. Metab.* **2018**, *73*, 234–240. [CrossRef] [PubMed]

33. Benítez-Porres, J.; Alvero-Cruz, J.R.; Sardinha, L.B.; López-Fernández, I.; Carnero, E.A. Valores de corte para clasificar niños y adolescentes activos utilizando el cuestionario de actividad física: PAQ-C y PAQ-A. *Nutr. Hosp.* **2016**, *33*, 1036–1044. [CrossRef]

34. Bernhardsen, G.P.; Stensrud, T.; Nystad, W.; Dalene, K.E.; Kolle, E.; Ekelund, U. Early life risk factors for childhood obesity-Does physical activity modify the associations? The MoBa cohort study. *Scand. J. Med. Sci. Sports* **2019**, *29*, 1636–1646. [CrossRef] [PubMed]

35. Chaput, J.P.; Lambert, M.; Mathieu, M.E.; Tremblay, M.S.; O'Loughlin, J.; Tremblay, A. Physical activity vs. sedentary time: Independent associations with adiposity in children. *Pediatr. Obes.* **2012**, *7*, 251–258. [CrossRef] [PubMed]

36. de Moraes Ferrari, G.L.; Oliveira, L.C.; Araujo, T.L.; Matsudo, V.; Barreira, T.V.; Tudor-Locke, C.; Katzmarzyk, P. Moderate-to-Vigorous Physical Activity and Sedentary Behavior: Independent Associations With Body Composition Variables in Brazilian Children. *Pediatr. Exerc. Sci.* **2015**, *27*, 380–389. [CrossRef]

37. Gába, A.; Mitáš, J.; Jakubec, L. Associations between accelerometer-measured physical activity and body fatness in school-aged children. *Environ. Health Prev. Med.* **2017**, *22*, 43. [CrossRef]

38. Hallal, P.C.; Reichert, F.F.; Clark, V.L.; Cordeira, K.L.; Menezes, A.M.B.; Eaton, S.; Ekelund, U.; Wells, J.C. Energy expenditure compared to physical activity measured by accelerometry and self-report in adolescents: A validation study. *PLoS ONE* **2013**, *8*, e77036. [CrossRef]

39. Herrmann, D.; Buck, C.; Sioen, I.; Kouride, Y.; Marild, S.; Molnár, D.; Mouratidou, T.; Pitsiladis, Y.; Russo, P.; Veidebaum, T.; et al. Impact of physical activity, sedentary behaviour and muscle strength on bone stiffness in 2-10-year-old children-cross-sectional results from the IDEFICS study. *Int. J. Behav. Nutr. Phys. Act.* **2015**, *12*, 112. [CrossRef]

40. Janz, K.F.; Boros, P.; Letuchy, E.M.; Kwon, S.; Burns, T.L.; Levy, S.M. Physical Activity, Not Sedentary Time, Predicts Dual-Energy X-ray Absorptiometry-measured Adiposity Age 5 to 19 Years. *Med. Sci. Sports Exerc.* **2017**, *49*, 2071–2077. [CrossRef]

41. Joensuu, L.; Syväoja, H.; Kallio, J.; Kulmala, J.; Kujala, U.M.; Tammelin, T.H. Objectively measured physical activity, body composition and physical fitness: Cross-sectional associations in 9- to 15-year-old children. *Eur. J. Sport Sci.* **2018**, *18*, 882–892. [CrossRef] [PubMed]

42. McCormack, L.; Meendering, J.; Specker, B.; Binkley, T. Associations Between Sedentary Time, Physical Activity, and Dual-Energy X-ray Absorptiometry Measures of Total Body, Android, and Gynoid Fat Mass in Children. *J. Clin. Densitom.* **2016**, *19*, 368–374. [CrossRef] [PubMed]

43. Santos-Magalhaes, A.F.; Aires, L.; Martins, C.; Silva, G.; Teixeira, A.M.; Mota, J.; Rama, L. Heart rate variability, adiposity, and physical activity in prepubescent children. *Clin. Auton. Res.* **2015**, *25*, 169–178. [CrossRef]

44. Sardinha, L.B.; Magalhães, J.P.; Santos, D.A.; Júdice, P.B. Sedentary patterns, physical activity, and cardiorespiratory fitness in association to glycemic control in type 2 diabetes patients. *Front. Physiol.* **2017**, *8*, 262. [CrossRef]

45. Amo-Setién, F.J.; Leal-Costa, C.; Abajas-Bustillo, R.; González-Lamuño, D.; Redondo-Figuero, C. Factors associated with grip strength among adolescents: An observational study. *J. Hand Ther.* **2020**, *33*, 96–102. [CrossRef] [PubMed]

46. Ainsworth, B.E.; Caspersen, C.J.; Matthews, C.E.; Mâsse, L.C.; Baranowski, T.; Zhu, W. Recommendations to improve the accuracy of estimates of physical activity derived from self report. *J. Phys. Act. Health* **2012**, *9* (Suppl. 1), S76–S84. [CrossRef]

47. Spruijt-Metz, D.; Wen, C.K.F.; Bell, B.M.; Intille, S.; Huang, J.S.; Baranowski, T. Advances and Controversies in Diet and Physical Activity Measurement in Youth. *Am. J. Prev. Med.* **2018**, *55*, e81–e91. [CrossRef]

48. Cain, K.L.; Sallis, J.F.; Conway, T.L.; Van Dyck, D.; Calhoon, L. Using accelerometers in youth physical activity studies: A review of methods. *J. Phys. Act. Health* **2013**, *10*, 437–450. [CrossRef]

49. Troiano, R.P.; Berrigan, D.; Dodd, K.W.; Mâsse, L.C.; Tilert, T.; McDowell, M. Physical activity in the United States measured by accelerometer. *Med. Sci. Sports Exerc.* **2008**, *40*, 181–188. [CrossRef]

50. Ellis, K.; Kerr, J.; Godbole, S.; Staudenmayer, J.; Lanckriet, G. Hip and Wrist Accelerometer Algorithms for Free-Living Behavior Classification. *Med. Sci. Sports Exerc.* **2016**, *48*, 933–940. [CrossRef]

51. Aadland, E.; Kvalheim, O.M.; Anderssen, S.A.; Resaland, G.K.; Andersen, L.B. The multivariate physical activity signature associated with metabolic health in children. *Int. J. Behav. Nutr. Phys. Act.* **2018**, *15*, 77. [CrossRef] [PubMed]

52. Robusto, K.M.; Trost, S.G. Comparison of three generations of ActiGraph™ activity monitors in children and adolescents. *J. Sports Sci.* **2012**, *30*, 1429–1435. [CrossRef] [PubMed]

53. Kozey, S.L.; Staudenmayer, J.W.; Troiano, R.P.; Freedson, P.S. Comparison of the ActiGraph 7164 and the ActiGraph GT1M during self-paced locomotion. *Med. Sci. Sports Exerc.* **2010**, *42*, 971–976. [CrossRef] [PubMed]

54. John, D.; Tyo, B.; Bassett, D.R. Comparison of four ActiGraph accelerometers during walking and running. *Med. Sci. Sports Exerc.* **2010**, *42*, 368–374. [CrossRef] [PubMed]

55. Grydeland, M.; Hansen, B.H.; Ried-Larsen, M.; Kolle, E.; Anderssen, S.A. Comparison of three generations of ActiGraph activity monitors under free-living conditions: Do they provide comparable assessments of overall physical activity in 9-year old children? *BMC Sports Sci. Med. Rehabil.* **2014**, *6*, 26. [CrossRef] [PubMed]

56. Robertson, W.; Stewart-Brown, S.; Wilcock, E.; Oldfield, M.; Thorogood, M. Utility of accelerometers to measure physical activity in children attending an obesity treatment intervention. *J. Obes.* **2011**, *2011*. [CrossRef]

57. Trost, S.G.; Mciver, K.L.; Pate, R.R. Conducting accelerometer-based activity assessments in field-based research. *Med. Sci. Sports Exerc.* **2005**, *37*, S531–S543. [CrossRef]

58. Aadland, E.; Ylvisåker, E. Reliability of the Actigraph GT3X+ Accelerometer in Adults under Free-Living Conditions. *PLoS ONE* **2015**, *10*, e0134606. [CrossRef]

59. Trost, S.G.; Ward, D.S.; Moorehead, S.M.; Watson, P.D.; Riner, W.; Burke, J.R. Validity of the computer science and applications (CSA) activity monitor in children. *Med. Sci. Sports Exerc.* **1998**, *30*, 629–633. [CrossRef]

60. O'Neil, M.E.; Fragala-Pinkham, M.A.; Forman, J.L.; Trost, S.G. Measuring reliability and validity of the ActiGraph GT3X accelerometer for children with cerebral palsy: A feasibility study. *J. Pediatr. Rehabil. Med.* **2014**, *7*, 233–240. [CrossRef]

61. Brocklebank, L.A.; Falconer, C.L.; Page, A.S.; Perry, R.; Cooper, A.R. Accelerometer-measured sedentary time and cardiometabolic biomarkers: A systematic review. *Prev. Med.* **2015**, *76*, 92–102. [CrossRef] [PubMed]

62. Rey Lopez, J.P.; Sabag, A.; Martinez Juan, M.; Rezende, L.F.M.; Pastor-Valero, M. Do vigorous-intensity and moderate-intensity physical activities reduce mortality to the same extent? A systematic review and meta-analysis. *BMJ Open Sport Exerc. Med.* **2020**, *6*, e000775. [CrossRef] [PubMed]

63. Malina, R.M.; Eisenmann, J.C.; Cumming, S.P.; Ribeiro, B.; Aroso, J. Maturity-associated variation in the growth and functional capacities of youth football (soccer) players 13–15 years. *Eur. J. Appl. Physiol.* **2004**, *91*, 555–562. [CrossRef]

64. Weber, D.R.; Leonard, M.B.; Zemel, B.S. Body composition analysis in the pediatric population. *Pediatr. Endocrinol. Rev.* **2012**, *10*, 130–139. [PubMed]

65. Kindblom, J.M.; Lorentzon, M.; Hellqvist, A.; Lönn, L.; Brandberg, J.; Nilsson, S.; Norjavaara, E.; Ohlsson, C. BMI changes during childhood and adolescence as predictors of amount of adult subcutaneous and visceral adipose tissue in men: The GOOD Study. *Diabetes* **2009**, *58*, 867–874. [CrossRef] [PubMed]

66. Strong, W.B.; Malina, R.M.; Blimkie, C.J.R.; Daniels, S.R.; Dishman, R.K.; Gutin, B.; Hergenroeder, A.C.; Must, A.; Nixon, P.A.; Pivarnik, J.M.; et al. Evidence based physical activity for school-age youth. *J. Pediatr.* **2005**, *146*, 732–737. [CrossRef]

67. Tremblay, M.S.; LeBlanc, A.G.; Kho, M.E.; Saunders, T.J.; Larouche, R.; Colley, R.C.; Goldfield, G.; Connor Gorber, S. Systematic review of sedentary behaviour and health indicators in school-aged children and youth. *Int. J. Behav. Nutr. Phys. Act.* **2011**, *8*, 98. [CrossRef]

68. Prentice-Dunn, H.; Prentice-Dunn, S. Physical activity, sedentary behavior, and childhood obesity: A review of cross-sectional studies. *Psychol. Health Med.* **2012**, *17*, 255–273. [CrossRef]

69. Gualdi-Russo, E.; Zaccagni, L.; Manzon, V.S.; Masotti, S.; Rinaldo, N.; Khyatti, M. Obesity and physical activity in children of immigrants. *Eur. J. Public Health* **2014**, *24*, 40–46. [CrossRef]

70. Draper, C.E.; Tomaz, S.A.; Stone, M.; Hinkley, T.; Jones, R.A.; Louw, J.; Twine, R.; Kahn, K.; Norris, S.A. Developing Intervention Strategies to Optimise Body Composition in Early Childhood in South Africa. *BioMed Res. Int.* **2017**, *2017*, 5283457. [CrossRef]

71. de Araújo, L.G.M.; Turi, B.C.; Locci, B.; Mesquita, C.A.A.; Fonsati, N.B.; Monteiro, H.L. Patterns of Physical Activity and Screen Time Among Brazilian Children. *J. Phys. Act. Health* **2018**, *15*, 457–461. [CrossRef] [PubMed]

72. Gerber, M.; Müller, I.; Walter, C.; du Randt, R.; Adams, L.; Gall, S.; Joubert, N.; Nqweniso, S.; Smith, D.; Steinmann, P.; et al. Physical activity and dual disease burden among South African primary schoolchildren from disadvantaged neighbourhoods. *Prev. Med.* **2018**, *112*, 104–110. [CrossRef]

73. Ferrari, G.L.; Kovalskys, I.; Fisberg, M.; Gómez, G.; Rigotti, A.; Sanabria, L.Y.C.; García, M.C.Y.; Torres, R.G.P.; Herrera-Cuenca, M.; Zimberg, I.Z.; et al. Comparison of self-report versus accelerometer—Measured physical activity and sedentary behaviors and their association with body composition in Latin American countries. *PLoS ONE* **2020**, *15*, e0232420. [CrossRef] [PubMed]

74. Guthold, R.; Stevens, G.A.; Riley, L.M.; Bull, F.C. Global trends in insufficient physical activity among adolescents: A pooled analysis of 298 population-based surveys with 1·6 million participants. *Lancet Child. Adolesc. Health* **2020**, *4*, 23–35. [CrossRef]

75. Christakoudi, S.; Tsilidis, K.K.; Muller, D.C.; Freisling, H.; Weiderpass, E.; Overvad, K.; Söderberg, S.; Häggström, C.; Pischon, T.; Dahm, C.C.; et al. A Body Shape Index (ABSI) achieves better mortality risk stratification than alternative indices of abdominal obesity: Results from a large European cohort. *Sci. Rep.* **2020**, *10*, 14541. [CrossRef]

76. Giudici, K.V.; Martini, L.A. Comparison between body mass index and a body shape index with adiponectin/leptin ratio and markers of glucose metabolism among adolescents. *Ann. Hum. Biol.* **2017**, *44*, 489–494. [CrossRef]

77. Freedman, D.S.; Sherry, B. The validity of BMI as an indicator of body fatness and risk among children. *Pediatrics* **2009**, *124* (Suppl. 1), S23–S34. [CrossRef]

78. Achamrah, N.; Colange, G.; Delay, J.; Rimbert, A.; Folope, V.; Petit, A.; Grigioni, S.; Déchelotte, P.; Coëffier, M. Comparison of body composition assessment by DXA and BIA according to the body mass index: A retrospective study on 3655 measures. *PLoS ONE* **2018**, *13*, e0200465. [CrossRef]

79. Lee, S.Y.; Ahn, S.; Kim, Y.J.; Ji, M.J.; Kim, K.M.; Choi, S.H.; Jang, H.C.; Lim, S. Comparison between Dual-Energy X-ray Absorptiometry and Bioelectrical Impedance Analyses for Accuracy in Measuring Whole Body Muscle Mass and Appendicular Skeletal Muscle Mass. *Nutrients* **2018**, *10*, 738. [CrossRef]

80. Alkahtani, S.A. A cross-sectional study on sarcopenia using different methods: Reference values for healthy Saudi young men. *BMC Musculoskelet. Disord.* **2017**, *18*, 119. [CrossRef]

81. Duz, S.; Kocak, M.; Korkusuz, F. Evaluation of body composition using three different methods compared to dual-energy X-ray absorptiometry. *Eur. J. Sport Sci.* **2009**, *9*, 181–190. [CrossRef]

82. Benito, P.J.; Gómez-Candela, C.; Cabañas, M.D.; Szendrei, B.; Castro, E.A. Comparison between different methods for measuring body fat after a weight loss program. *Rev. Bras. Med. Esporte* **2019**, *25*, 474–479. [CrossRef]

83. Must, A.; Tybor, D.J. Physical activity and sedentary behavior: A review of longitudinal studies of weight and adiposity in youth. *Int. J. Obes.* **2005**, *29* (Suppl. 2), S84–S96. [CrossRef]

84. Hong, H.-R.; Ha, C.-D.; Kong, J.-Y.; Lee, S.-H.; Song, M.-G.; Kang, H.-S. Roles of physical activity and cardiorespiratory fitness on sex difference in insulin resistance in late elementary years. *J. Exerc. Nutr. Biochem.* **2014**, *18*, 361–369. [CrossRef] [PubMed]

85. Ishikawa-Takata, K.; Kaneko, K.; Koizumi, K.; Ito, C. Comparison of physical activity energy expenditure in Japanese adolescents assessed by EW4800P triaxial accelerometry and the doubly labelled water method. *Br. J. Nutr.* **2013**, *110*, 1347–1355. [CrossRef] [PubMed]

86. Lv, Y.; Cai, L.; Gui, Z.; Zeng, X.; Tan, M.; Wan, N.; Lai, L.; Lu, S.; Tan, W.; Chen, Y. Effects of physical activity and sedentary behaviour on cardiometabolic risk factors and cognitive function in children: Protocol for a cohort study. *BMJ Open* **2019**, *9*, e030322. [CrossRef] [PubMed]

87. Migueles, J.H.; Cadenas-Sanchez, C.; Ekelund, U.; Delisle Nyström, C.; Mora-Gonzalez, J.; Löf, M.; Labayen, I.; Ruiz, J.R.; Ortega, F.B. Accelerometer Data Collection and Processing Criteria to Assess Physical Activity and Other Outcomes: A Systematic Review and Practical Considerations. *Sports Med.* **2017**, *47*, 1821–1845. [CrossRef]

88. Matthews, C.E.; Ainsworth, B.E.; Thompson, R.W.; Bassett, D.R.J. Sources of variance in daily physical activity levels as measured by an accelerometer. *Med. Sci. Sports Exerc.* **2002**, *34*, 1376–1381. [CrossRef]

Sustainability **2021**, *13*, 335

Publisher's Note: MDPI stays neutral with regard to jurisdictional claims in published maps and institutional affiliations.